Laboratory Manual
for
Modern Welding

by

William A. Bowditch
Technical-Vocational Education Consultant
Gaylord, Michigan
Life Member of the American Welding Society
Member of the American Vocational Association

and

Kevin E. Bowditch
Welding Engineer Specialist
Subaru-Isuzu Automotive Inc.
Lafayette, Indiana
Member of the American Welding Society
Member of the American Vocational Association

Publisher
The Goodheart-Willcox Company, Inc.
Tinley Park, Illinois

Copyright 1997

by

THE GOODHEART-WILLCOX COMPANY, INC.

Previous Editions Copyright 1989, 1994

All rights reserved. No part of this book may be reproduced, stored in a retrieval system, or transmitted in any form or by any means, electronic, mechanical, photocopying, recording, or otherwise, without the prior written permission of The Goodheart-Willcox Company, Inc. Manufactured in the United States of America.

International Standard Book Number 1-56637-331-X
3 4 5 6 7 8 9 10 97 00 99

INTRODUCTION

The **Laboratory Manual** for **Modern Welding** is intended to be used with the **Modern Welding** text. This manual will help you to practice the welding techniques for the various welding processes presented in the text. Answering questions in the various lessons will help you to ensure that you have mastered the technical knowledge presented in the text.

Safety is a very important aspect of welding and is thoroughly discussed in Chapter 1. Safety notes and cautions are also printed in red throughout the textbook wherever they apply. Prior to beginning any new welding process, you should read about the process in the text and then satisfactorily pass the safety test(s) for that process.

Modern Welding and the **Laboratory Manual** for **Modern Welding** are divided into nine parts. Each part covers one or more welding processes. Part 1 covers safety, print reading, reading welding symbols, and welding and cutting processes. Part 2 covers oxyfuel gas welding and cutting, soldering, brazing, and braze welding. Part 3 covers the shielded metal arc welding (SMAW) processes. Part 4 covers the gas metal arc welding (GMAW) and gas tungsten arc welding (GTAW) processes. Part 5 explains the arc cutting and oxygen arc cutting and gouging processes. Part 6 covers various resistance welding processes and special processes of welding and cutting are explained in Part 7. Metal Technology, Part 8, covers the production, properties, and heat treatment of metals. Weld inspection and testing, weld and welder qualifications, and the welding shop are discussed in Part 9, Professional Welding.

Although your instructor may assign chapters out of sequence, it is important that you study Part 1 first. Part 1 (Chapters 1-4) should be studied first because it covers print reading and reading AWS welding symbols. All the Jobs in this manual are drawn using AWS symbols. Therefore, you must know how to read these drawings and welding symbols.

Each lesson includes objectives, section and heading references, and specific instructions. Several types of questions are used, including matching, true or false, identification of parts, multiple choice, fill-in, and written essay answers. Many of the questions involve math calculations while others require the use of tables and charts. The questions are designed to test your technical knowledge of welding techniques. Do your best to answer these questions completely and accurately.

Most of the lessons include one or more Jobs. Read the Job directions carefully. Be certain to follow all of the safety precautions. Inspection criteria and weld test procedures will be given for each Job. Inspect each weld prior to submitting it to the instructor for grading.

It is strongly advised that you keep a record of your lesson and job grades on the "Lesson, Job, and Performance Objective Record" on pages 7–13 of this manual. Besides keeping track of grades and the completion of lab assignments, the completed chart can be included with your resume to demonstrate your experience and knowledge in welding to an employer.

William A. Bowditch
Kevin E. Bowditch

TABLE OF CONTENTS

PART 1
Welding Fundamentals

Lesson 1A	General Shop Safety Test	15
Lesson 1B	Oxyfuel Gas Welding and Cutting Safety Test	17
Lesson 1C	Arc Welding and Cutting Safety Test	21
Lesson 1D	Resistance and Other Welding Processes Safety Test	23
Lesson 2	Print Reading	25
Lesson 3	Reading Welding Symbols	29
Lesson 4A	Welding and Cutting Processes	37
Lesson 4B	Welding and Cutting Processes	41
Lesson 4C	Welding and Cutting Processes	45

PART 2
Oxyfuel Gas Processes

Lesson 5A	Oxyfuel Gas Welding Equipment and Supplies	49
Lesson 5B	Oxyfuel Gas Welding and Cutting Safety	53
Lesson 6A	Oxyfuel Gas Welding—Turning the Outfit On and Off	57
Lesson 6B	Oxyfuel Gas Welding—Running a Continuous Weld Pool	63
Lesson 6C	Oxyfuel Gas Welding—Welding Mild Steel in the Flat Welding Position	67
Lesson 6D	Oxyfuel Gas Welding—Welding Mild Steel in the Horizontal Welding Position	77
Lesson 6E	Oxyfuel Gas Welding—Welding Mild Steel in the Vertical Welding Position	83

Lesson 6F	Oxyfuel Gas Welding—Welding Mild Steel in the Overhead Welding Position	89
Lesson 7	Oxyfuel Gas Cutting Equipment and Supplies	93
Lesson 8	Oxyfuel Gas Cutting—Cutting Steel	97
Lesson 9	Soldering	107
Lesson 10A	Brazing and Braze Welding	113
Lesson 10B	Brazing and Braze Welding	117

PART 3
Shielded Metal Arc Welding

Lesson 11A	Shielded Metal Arc Welding Equipment and Supplies	121
Lesson 11B	Shielded Metal Arc Welding Equipment and Supplies	125
Lesson 11C	Arc Safety Welding	129
Lesson 12A	Arc Welding Fundamentals	133
Lesson 12B	Arc Welding in the Flat Welding Position	145
Lesson 12C	Arc Welding Out of Position	153
Lesson 12D	Arc Welding Out of Position	159

PART 4
Gas Tungsten and Gas Metal Arc Welding

Lesson 13A	GTAW Equipment and Supplies	167
Lesson 13B	GTAW Equipment and Supplies	171
Lesson 14A	Gas Tungsten Arc Welding Principles	175
Lesson 14B	Gas Tungsten Arc Welding	187
Lesson 15A	GMAW Equipment and Supplies	201
Lesson 15B	GMAW Equipment and Supplies	205
Lesson 15C	GMAW and FCAW Procedures	213
Lesson 15D	GMAW and FCAW Procedures	229
Lesson 15E	GMA Spot Welding	243

PART 5
Arc Cutting

Lesson 16A	Arc Cutting and Gouging Safety	251
Lesson 16B	Arc Cutting and Gouging Equipment and Supplies	255
Lesson 17	Arc Cutting and Gouging	259

PART 6
Resistance Welding

Lesson 18A	Resistance Welding Safety	269
Lesson 18B	Resistance Welding Machines	271
Lesson 18C	Resistance Welding Electrical Components	277
Lesson 19	Set-Up and Welding	283

PART 7
Special Processes

Lesson 20A	Arc-Related Welding Processes	295
Lesson 20B	Solid-State and Other Welding Processes	299
Lesson 21	Ferrous Welding Applications	303
Lesson 22	Nonferrous Welding Applications	313
Lesson 23	Pipe and Tube Welding	319
Lesson 24	Special Cutting Processes	329
Lesson 25	Automatic and Robotic Welding	333
Lesson 26	Metal Surfacing	337

PART 8
Metal Technology

Lesson 27	Production of Metals	343
Lesson 28	Metal Properties and Identification	347
Lesson 29	Heat Treatment of Metals	353

PART 9
Professional Welding

Lesson 30	Inspecting and Testing Welds	359
Lesson 31	Procedure and Welder Qualifications	371
Lesson 32	The Welding Shop	379
Lesson 33	Getting and Holding a Job in the Welding Industry	381

LESSON, JOB, AND PERFORMANCE OBJECTIVE RECORD

Name_____ Course_____ Class_____

School_____ Instructor_____

As stated in the introduction, the following chart has been provided to assist you in keeping accurate records of both your grades and the completion of the lab assignments. Grades should be promptly and neatly recorded for future reference.

Grade

PART 1
Welding Fundamentals

Lesson 1A	General Shop Safety Test	_____
Lesson 1B	Oxyfuel Gas Welding and Cutting Safety Test	_____
Lesson 1C	Arc Welding and Cutting Safety Test	_____
Lesson 1D	Resistance and Other Welding Processes Safety Test	_____
Lesson 2	Print Reading	_____
Lesson 3	Reading Welding Symbols	_____
Lesson 4A	Welding and Cutting Processes	_____
Lesson 4B	Welding and Cutting Processes	_____
Lesson 4C	Welding and Cutting Processes	_____

PART 2
Oxyfuel Gas Processes

Lesson 5A	Oxyfuel Gas Welding Equipment and Supplies	_____
Lesson 5B	Oxyfuel Gas Welding and Cutting Safety	_____
Lesson 6A	Oxyfuel Gas Welding—Turning the Outfit On and Off	_____
Job 6A-1	Turning On, Lighting, and Shutting Down the Oxyacetylene Welding Outfit	_____
Lesson 6B	Oxyfuel Gas Welding—Running a Continuous Weld Pool	_____
Job 6B-1	Running a Continuous Weld Pool	_____
Job 6B-2	Square-Groove Weld on an Outside Corner	_____
Lesson 6C	Oxyfuel Gas Welding—Welding Mild Steel in the Flat Welding Position	_____
Job 6C-1	Fillet Weld on a Lap Joint in the Flat Welding Position	_____
Job 6C-2	Fillet Weld on an Inside Corner Joint in the Flat Welding Position	_____

Copyright Goodheart-Willcox Co., Inc.

		Grade
Job 6C-3	Square-Groove Weld on an Outside Corner in the Flat Welding Position	_____
Job 6C-4	Square-Groove Weld on a Butt Joint in the Flat Welding Position	_____
Job 6C-5	Fillet Weld on an Inside Corner and a V-Groove Weld on an Outside Corner in the Flat Welding Position	_____
Job 6C-6	Fillet Weld on a Lap Joint in the Flat Welding Position (thicker metal)	_____
Job 6C-7	V-Groove Weld on a Butt Joint in the Flat Welding Position (thicker metal)	_____
Lesson 6D	Oxyfuel Gas Welding—Welding Mild Steel in the Horizontal Welding Position	_____
Job 6D-1	Fillet Weld on a Lap Joint in the Horizontal Welding Position	_____
Job 6D-2	Fillet Weld on a T-Joint in the Horizontal Welding Position	_____
Job 6D-3	Square-Groove Weld on a Butt Joint in the Horizontal Welding Position	_____
Lesson 6E	Oxyfuel Gas Welding—Welding Mild Steel in the Vertical Welding Position	_____
Job 6E-1	Fillet Weld on a Lap Joint in the Vertical Welding Position	_____
Job 6E-2	Fillet Weld on a T-Joint in the Vertical Welding Position	_____
Job 6E-3	V-Groove Weld on an Outside Corner in the Vertical Welding Position	_____
Lesson 6F	Oxyfuel Gas Welding—Welding Mild Steel in the Overhead Welding Position	_____
Job 6F-1	Square-Groove Weld on a Butt Joint, a Fillet Weld on a Lap Joint, and a Fillet Weld on a T-Joint in the Overhead Welding Position	_____
Lesson 7	Oxyfuel Gas Cutting Equipment and Supplies	_____
Lesson 8	Oxyfuel Gas Cutting—Cutting Steel	_____
Job 8-1	Cutting Mild Steel with the Oxyfuel Gas Torch	_____
Job 8-2	Cutting Shapes with the Oxyfuel Gas Torch	_____
Job 8-3	Removing Weld Reinforcement from the Face of a Weld	_____
Job 8-4	Setting Up and Inspecting Oxyfuel Gas Track Burner	_____
Job 8-5	Making Straight-Line Cuts on Steel to Produce Square and Beveled Edges	_____
Lesson 9	Soldering	_____
Job 9-1	Soldering Copper Fittings Used in Plumbing	_____
Job 9-2	Soldering a Folded Seam	_____

		Grade
Lesson 10A	Brazing and Braze Welding	____
Job 10A-1	Braze Welding a Butt Joint, a Lap Joint, and a T-Joint in the Flat Welding Position	____
Lesson 10B	Brazing and Braze Welding	____
Job 10B-1	Braze Welding a V-Groove on an Outside Corner in the Flat Welding Position	____

PART 3
Shielded Metal Arc Welding

Lesson 11A	Shielded Metal Arc Welding Equipment and Supplies	____
Lesson 11B	Shielded Metal Arc Welding Equipment and Supplies	____
Lesson 11C	Arc Welding Safety	____
Lesson 12A	Arc Welding Fundamentals	____
Job 12A-1	Inspecting and Setting Up the Arc Welding Station	____
Job 12A-2	Striking an Arc and Running Short Beads	____
Job 12A-3	Running Arc Beads	____
Job 12A-4	Running Weaving Beads with E6012 Electrodes	____
Job 12A-5	Running Stringer Beads with E6010, E6011, E6013, and Other Electrodes	____
Lesson 12B	Arc Welding in the Flat Welding Position	____
Job 12B-1	Square-Groove Weld on an Edge Joint in the Flat Welding Position	____
Job 12B-2	Fillet Weld on a Lap Joint in the Flat Welding Position	____
Job 12B-3	Fillet Weld on a T-Joint in the Flat Welding Position	____
Job 12B-4	Bevel-Groove Weld on a Butt Joint in the Flat Welding Position	____
Lesson 12C	Arc Welding Out of Position	____
Job 12C-1	Fillet Weld on a Lap Joint in the Horizontal Welding Position	____
Job 12C-2	Fillet Weld on a T-Joint in the Horizontal Welding Position	____
Job 12C-3	Bevel-Groove Weld on a Butt Joint and a V-Groove Weld on an Outside Corner Joint in the Horizontal Welding Position	____
Lesson 12D	Arc Welding Out of Position	____
Job 12D-1	Fillet Weld on a Lap Joint in the Vertical Welding Position	____
Job 12D-2	Fillet Weld on a T-Joint in the Vertical Welding Position	____
Job 12D-3	V-Groove Weld on a Butt Joint in the Vertical Welding Position	____
Job 12D-4	Fillet Weld on a Lap Joint in the Overhead Welding Position	____
Job 12D-5	Fillet Weld on a T-Joint in the Overhead Welding Position	____

Grade

Job 12D-6	Bevel-Groove Weld on a Butt Joint in the Overhead Welding Position	_____

PART 4
Gas Tungsten and Gas Metal Arc Welding

Lesson 13A	GTAW Equipment and Supplies	_____
Lesson 13B	GTAW Equipment and Supplies	_____
Lesson 14A	Gas Tungsten Arc Welding Principles	_____
Job 14A-1	Fillet Weld on a Lap Joint in the Flat Welding Position	_____
Job 14A-2	Fillet Weld on an Inside Corner in the Flat Welding Position	_____
Job 14A-3	Square-Groove Weld on a Butt Joint in the Flat Welding Position	_____
Lesson 14B	Gas Tungsten Arc Welding	_____
Job 14B-1	Fillet Weld on a Lap Joint in the Horizontal Welding Position	_____
Job 14B-2	Fillet Weld on an Inside Corner and a T-Joint in the Horizontal Welding Position	_____
Job 14B-3	Square-Groove Weld on a Butt Joint in the Horizontal Welding Position	_____
Job 14B-4	Fillet Weld on a Lap Joint in the Vertical Welding Position	_____
Job 14B-5	Fillet Weld on a T-Joint in the Vertical Welding Position	_____
Job 14B-6	Square-Groove Weld on a Butt Joint in the Vertical Welding Position	_____
Job 14B-7	Fillet Weld on a Lap Joint in the Overhead Welding Position	_____
Job 14B-8	Fillet Weld on a T-Joint in the Overhead Welding Position	_____
Job 14B-9	Square-Groove Weld on a Butt Joint in the Overhead Welding Position	_____
Job 14B-10	Groove Welds on Stainless Steel, Using the GTAW Process	_____
Lesson 15A	GMAW Equipment and Supplies	_____
Lesson 15B	GMAW Equipment and Supplies	_____
Job 15B-1	Adjusting the GMA and FCA Welding Machine	_____
Lesson 15C	GMAW and FCAW Procedures	_____
Job 15C-1	Setting the GMA Welding Machine and Making a Fillet Weld on a Lap Joint in the Flat Welding Position	_____
Job 15C-2	Fillet Weld on an Inside Corner and a T-Joint in the Flat Welding Position	_____
Job 15C-3	Square-Groove Weld on a Butt Joint in the Flat Welding Position	_____
Job 15C-4	Fillet Weld on a Lap Joint in the Horizontal Welding Position	_____
Job 15C-5	Fillet Weld on a T-Joint in the Horizontal Welding Position	_____

		Grade
Job 15C-6	Square-Groove Weld on a Butt Joint in the Horizontal Welding Position	____
Lesson 15D	GMAW and FCAW Procedures	____
Job 15D-1	Fillet Weld on a Lap Joint in the Vertical Welding Position	____
Job 15D-2	Fillet Weld on an Inside Corner and a T-Joint in the Vertical Welding Position	____
Job 15D-3	Bevel-Groove Weld on a Butt Joint in the Vertical Welding Position	____
Job 15D-4	Fillet Weld on a Lap Joint in the Overhead Welding Position	____
Job 15D-5	Fillet Weld on a T-Joint in the Overhead Welding Position	____
Job 15D-6	Square-Groove Weld on a Butt Joint in the Overhead Welding Position	____
Lesson 15E	GMA Spot Welding	____
Job 15E-1	Setting the GMAW Machine for Spot Welding and Making Spot Welds	____

PART 5
Arc Cutting

Lesson 16A	Arc Cutting and Gouging Safety	____
Lesson 16B	Arc Cutting Equipment and Supplies	____
Lesson 17	Arc and Oxygen Arc Cutting	____
Job 17-1	Inspecting an Air Carbon Arc Welding Station	____
Job 17-2	Cutting and Piercing with the Air Carbon Arc Process	____
Job 17-3	Removing a Weld or Weld Reinforcement Using the Air Carbon Arc Process	____
Job 17-4	Piercing and Cutting Using the Plasma Arc Cutting Process	____
Job 17-5	Cutting a Shape Using the Plasma Arc Cutting Process	____

PART 6
Resistance Welding

Lesson 18A	Resistance Welding Safety	____
Lesson 18B	Resistance Welding Machines	____
Job 18B-1	Major Components of the Resistance Welding Machines in Your Shop	____
Lesson 18C	Resistance Welding Electrical Components	____

		Grade
Job 18C-1	Machine Cooling and Electrodes	_____
Lesson 19	Set-Up and Welding	_____
Job 19-1	Setting the Variables and Making Spot Welds	_____
Job 19-2	Setting the Variables and Making Spot Welds	_____

PART 7
Special Processes

Lesson 20A	Arc-Related Welding Processes	_____
Lesson 20B	Solid-State and Other Welding Processes	_____
Lesson 21	Ferrous Welding Applications	_____
Job 21-1	Square-Groove Weld on a Butt Joint on Stainless Steel Using SMAW in the Flat Welding Position	_____
Job 21-2	Fillet Weld on a T-Joint on Stainless Steel Using SMAW in the Flat Welding Position	_____
Job 21-3	Square-Groove Weld on a Butt Joint and a Fillet Weld in the Flat Welding Position	_____
Job 21-4	Square or V-Groove Butt Weld on Cast Iron Using SMAW	_____
Lesson 22	Nonferrous Welding Applications	_____
Job 22-1	Butt Welds and Fillet Welds on Aluminum Using the GTAW Process	_____
Lesson 23	Pipe and Tube Welding	_____
Job 23-1	Welding Mild Steel Pipe in the 1G Position Using the SMAW Process	_____
Job 23-2	Welding Mild Steel Pipe in the 2G and 5G Positions Using the SMAW Process	_____
Job 23-3	Welding Mild Steel Pipe in the 5G Position Using the GTAW Process	_____
Job 23-4	Welding Mild Steel Pipe in the 2G Position Using the GMAW Process	_____
Job 23-5	Welding Mild Steel Pipe in the 6G Position Using the GMAW Process	_____
Lesson 24	Special Cutting Processes	_____
Lesson 25	Automatic and Robotic Welding	_____
Lesson 26	Metal Surfacing	_____
Job 26-1	Hardfacing a Steel Plate Using the Oxyfuel Gas Process	_____

Name: _____ Lesson, Job, and Performance Objective Record

Grade

PART 8
Metal Technology

Lesson 27	Production of Metals	_____
Lesson 28	Metal Properties and Identification	_____
Job 28-1	Identification of Metals	_____
Lesson 29	Heat Treatment of Metals	_____
Job 29-1	Heat Treating a Cold Chisel	_____

PART 9
Professional Welding

Lesson 30	Inspecting and Testing Welds	_____
Job 30-1	Magnetic Particle Inspection	_____
Job 30-2	Liquid Penetrant Inspection	_____
Job 30-3	Guided Bend Tests	_____
Job 30-4	Tensile Testing	_____
Job 30-5	Hardness Testing	_____
Lesson 31	Procedure and Welder Qualifications	_____
Job 31-1	Writing a Welding Procedure Specification and Qualifying the WPS	_____
Job 31-2	Performing a Fillet Performance Test	_____
Lesson 32	The Welding Shop	_____
Lesson 33	Getting and Holding a Job in the Welding Industry	_____
Job 33-1	Completing a Job Application	_____

Copyright Goodheart-Willcox Co., Inc.

Lesson 1A

GENERAL SHOP SAFETY

Name _____ Date _____ Lesson Grade _____
Class _____ Instructor _____

LEARNING OBJECTIVE
✶ You will able to identify several of the safety hazards and precautions required when working in any typical shop area.

INSTRUCTIONS
Carefully read Headings 1.1 through 1.3 and Figures 1-1 and 1-2 in the text. Then answer the following questions.

1. *True or False?* Your age, state of health, job skills, and attitude have little to do with how you feel about yourself or your job.

2. *True or False?* Health is not considered a physical factor in accidents.

3. *True or False?* Paint, oil, welding rods, and cleaning chemicals should be stored in a steel cabinet for safety.

4. *True or False?* A person who is under stress may be distracted by thoughts of love, worry, hate, illness, or fatigue.

5. *True or False?* Fire extinguishers and fire blankets are mounted on a surface that is painted a bright yellow.

6. *True or False?* It is good safety practice when working around electricity to clearly mark voltages of 220V or higher with danger signs.

7. *True or False?* When lifting, it is a good practice to wear a back brace, lift with your legs, and keep your back straight.

8. *True or False?* All hand tools should be inspected for size, worn power cords, and loose parts.

9. *True or False?* Safety hazards, locations of fire extinguishers, the proper placement of hands and feet, and machine safety features should be covered during job or task training.

10. *True or False?* Thinking about pleasant things to pass the time will help to promote safety in the shop.

1. _____
2. _____
3. _____
4. _____
5. _____
6. _____
7. _____
8. _____
9. _____
10. _____

11. *True or False?* Engaging in horseplay in the shop may be considered a lack of wisdom or intelligence.

11. _____

12. *True or False?* Fire extinguisher locations are painted yellow and black.

12. _____

13. _____ is considered an important factor in shop safety.
    A. Housekeeping
    B. A good attitude
    C. The proper storage of hazardous materials
    D. Adequate job training
    E. All of the above.

13. _____

14. Which of the following line voltages need not be marked *danger* or *danger high voltage?*
    A. 24 volts
    B. 115 volts
    C. 230 volts
    D. 220 volts
    E. 440 volts

14. _____

15. _____ might cause a person to be under stress.
    A. Love
    B. Worry
    C. Hate
    D. Illness
    E. All of the above.

15. _____

16. _____ are considered hazardous machines.
    A. Punch presses
    B. Metal sheers
    C. Pedestal grinders
    D. Cut-off saws
    E. All of the above.

16. _____

17. _____ may lead to a back injury when lifting.
    A. Wearing a back brace
    B. Bending your back slightly
    C. Lifting with your legs
    D. Lifting over 35 pounds
    E. None of the above.

17. _____

18. Well-designed welding shops have _____ ceilings to improve ventilation.

18. _____

19. Many companies require a worker to wear a _____ when lifting anything.

19. _____

20. Dangerous corners or stairway overhangs should be painted with alternating yellow and _____ angled stripes for high visibility.

20. _____

Lesson 1B

OXYFUEL GAS WELDING AND CUTTING SAFETY TEST

Name _____ Date _____ Lesson Grade _____
Class _____ Instructor _____

LEARNING OBJECTIVE
* You will be able to list several of the hazards and safety precautions to observe when working with oxyfuel gas welding and cutting equipment.

INSTRUCTIONS
Carefully read Heading 1.4. Also study Figures 1-4 through 1-10 in the text. Then answer the following questions.

1. The filter lens recommended to be worn when doing most oxyfuel gas welding or cutting is a number _____.
   A. 1–3
   B. 4–6
   C. 7–9
   D. 10–12
   E. 13–15

   1. _____

2. The recommended fabric for clothing worn while oxyfuel gas welding or cutting is _____.
   A. wool
   B. flame-retardant-treated cotton
   C. leather
   D. All the above.
   E. None of the above.

   2. _____

3. Metal that has been left somewhere to be cooled should be clearly marked _____.
   A. Do not touch
   B. HOT
   C. HOT metal do not touch
   D. Touch at your own risk
   E. Keep away—HOT

   3. _____

4. Acetylene gas should never be used at a pressure above _____.
   A. 10 psig
   B. 14 psig
   C. 14.7 psig
   D. 15 psig
   E. 32 psig

   4. _____

Copyright Goodheart-Willcox Co., Inc.

17

5. The oxygen cylinder valve should be opened _____ when the oxygen cylinder is in use.
   A. all the way out
   B. to the position for the correct pressure
   C. one full turn
   D. 1/4 turn so it can be turned off quickly
   E. 1/2 turn

   5. _____

6. If the highest pressure reading on the face of a gauge is 200 psig, what is the highest working pressure at which this gauge should be used?
   A. 50 psig
   B. 100 psig
   C. 150 psig
   D. 175 psig
   E. 200 psig

   6. _____

7. _____ may form toxic or deadly fumes when heated or welding.
   A. Beryllium
   B. Chromium
   C. Cadmium
   D. Lead
   E. All of the above.

   7. _____

8. When using either oxygen or acetylene cylinders, cylinders should always be _____.
   A. fastened to a wall or column
   B. moved with the safety cap in place
   C. stored upright in a special room or cage
   D. secured with sturdy chains or steel bands
   E. All of the above.

   8. _____

9. *True or False?* Shirts with open pockets and trousers without cuffs should be worn when welding.

   9. _____

10. *True or False?* Leather gauntlet-type gloves are best for oxyfuel gas welding in the overhead position.

    10. _____

11. *True or False?* Ultraviolet and infrared rays are created when oxyfuel gas welding.

    11. _____

12. *True or False?* The oxyfuel gas flame should be lighted with a safety match.

    12. _____

13. *True or False?* The American Welding Society recommends that a label be attached to the major equipment in an oxyfuel gas welding station to warn of potential hazards.

    13. _____

14. *True or False?* All gases should be bled from an oxyfuel gas welding station when the station is shut down.

    14. _____

15. *True or False?* Whenever the welding torch is not in the welder's hand, it must be turned off.

    15. _____

16. *True or False?* When oxyfuel gas cutting, the area must be cleared of combustible materials and at least one fire watch posted with a fire extinguisher.

    16. _____

Copyright Goodheart-Willcox Co., Inc.

17. *True or False?* Sodium cyanide compounds can create toxic or deadly fumes when welded.

18. To avoid possible burns, _____ should be used to pick up hot metal in the welding shop.

19. The pressure regulator on a welding station is OFF when the regulator's adjusting screw is turned all the way _____. (in/out)

20. _____ is *not* normally worn when oxyfuel gas welding or cutting.
    A. A face shield with filter quality plastic
    B. Welding goggles with round lenses
    C. Welding goggles with a rectangular lens
    D. A welding helmet

17. _____

18. _____

19. _____

20. _____

Lesson 1C

ARC WELDING AND CUTTING SAFETY

Name _____ Date _____ Lesson Grade _____
Class _____ Instructor _____

> **LEARNING OBJECTIVE**
>
> ✶ You will be able to list several hazards and safety precautions to observe when working with arc welding and cutting equipment.
>
> **INSTRUCTIONS**
> Read carefully Heading 1.5 and study Figure 1-11 in the text. Then answer the following questions.

1. *True or False?* Ultraviolet and infrared rays are created when arc welding or cutting.

2. *True or False?* High-pressure gas cylinders can be moved using a cylinder truck or by slightly tipping the cylinder and rolling it on its edge.

3. *True or False?* The American Welding Society recommends that a label be attached to all major equipment in an arc welding or cutting station to warn of dangers.

4. *True or False?* It is recommended that the arc welding station inspection be made with the power on.

5. *True or False?* Toxic fumes given off while arc welding should be removed at a point above the welder's head.

6. *True or False?* Cadmium, chromium, beryllium, zinc, and lead can give off toxic or deadly fumes when welded or cut.

7. *True or False?* It is recommended that a filter-type respirator be worn when welding or cutting on materials that may produce toxic or poisonous fumes.

8. When using or handling high pressure cylinders, _____ is *not* recommended.
   A. always installing the safety cap when not in use
   B. storing the cylinders separately in a special cage or room
   C. using chains or steel bands to secure the cylinders
   D. laying the cylinder on the floor and rolling it
   E. using a cylinder truck to move the cylinder

1. _____
2. _____
3. _____
4. _____
5. _____
6. _____
7. _____
8. _____

Copyright Goodheart-Willcox Co., Inc.

9. When arc cutting, it is recommended that a fire watch be posted with a _____.

9. _____

10. When arc welding or cutting and it is necessary to run the welding leads across an aisle or high-traffic area, you should _____.
    A. post a warning sign
    B. post a detour sign
    C. lay the leads under a piece of channel iron
    D. lay the leads under a piece of angle iron
    E. post a sign warning that the leads are in the aisle

10. _____

Lesson 1D

RESISTANCE AND SPECIAL WELDING PROCESSES SAFETY

Name _____ Date _____ Lesson Grade _____
Class _____ Instructor _____

LEARNING OBJECTIVE
★ You will be able to identify several of the hazards and safety precautions to observe when working with resistance and some other welding processes.

INSTRUCTIONS
Carefully read Headings 1.6 through 1.8. Then answer the following questions.

1. *True or False?* An unbreakable plastic face shield can be worn while resistance spot or seam welding.

2. *True or False?* It is *not* required for the welder to wear gloves while resistance welding.

3. *True or False?* You should not carry matches, butane lighters, or plastic combs in your pockets when resistance welding.

4. *True or False?* Hot sparks and molten metal may fly about when resistance spot welding.

5. *True or False?* While resistance spot welding, it is *not* a good safety practice to stand on an uninsulated platform.

6. *True or False?* Explosive welding requires special training.

7. *True or False?* Explosive welding and plasma welding are two welding processes that produce sounds so loud that they could damage the welder's hearing.

8. Which of the following welding processes require(s) special filter lenses to filter more than ultraviolet and infrared rays?
    A. Laser beam welding
    B. Gas tungsten arc welding
    C. Gas metal arc welding
    D. Electron beam welding
    E. Resistance projection welding

1. _____
2. _____
3. _____
4. _____
5. _____
6. _____
7. _____
8. _____

9. _____ light rays are dangerous light rays created while oxyfuel gas welding.
   A. Strobe
   B. Ultraviolet
   C. Incandescent
   D. Gaseous
   E. Infrared

9. _____

10. Why is it dangerous to stand within a robot's working volume? _____
_____
_____
_____

Lesson 2

PRINT READING

Name _____ Date _____ Lesson Grade _____
Class _____ Instructor _____

LEARNING OBJECTIVE
* You will be able to identify the various views in a sketch or drawing. You will also be able to follow a line or point from one view to another and find the sizes of various parts from a print.

INSTRUCTIONS
Carefully read Headings 2.1 through 2.7 of the text. Also study Figures 2-1 through 2-7. Then answer the following questions.

Use the drawing below when answering questions 1–10.

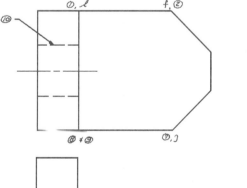
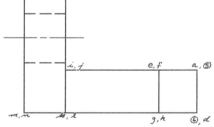
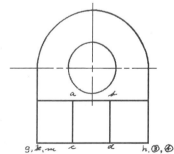

Identify the ends of the lines shown on the drawing from lines already identified in other views. In question 1, identify the missing point 1. In question 2, identify point 2. Remember, when a line (s-t) appears as a point, it is identified as (s,t). The first letter written is the point closest to you: "s" is closest and "t" is farthest away.

Copyright Goodheart-Willcox Co., Inc.

25

26   Modern Welding Lab Manual

1. _____
2. _____
3. _____
4. _____
5. _____
6. _____
7. _____
8. _____
9. _____
10. What do these hidden lines represent? _____

Use the drawing below when answering questions 11-20.

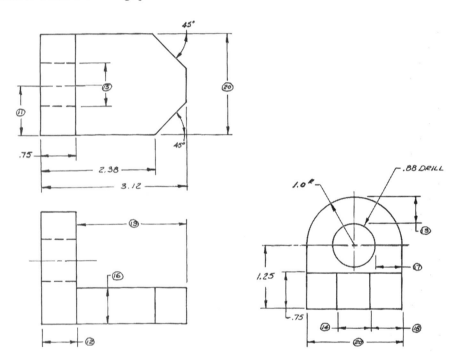

What is the distance between arrowheads or the size of the following numbers on the drawing?

11. _____
12. _____
13. _____
14. _____
15. _____
16. _____
17. _____
18. _____
19. _____
20. _____

Name: _____ Lesson 2 Print Reading 27

The views in the drawing on the previous page are lettered A, B, and C.

21. What view is view "A"? _____

22. What view is view "B"? _____

23. What view is view "C"? _____

24. What method of projection was used on the drawings? _____

25. Are these drawings sketches or mechanical drawings? _____

Lesson 3
READING WELDING SYMBOLS

Name _____ Date _____ Lesson Grade _____
Class _____ Instructor _____

LEARNING OBJECTIVE
✶ You will be able to identify the basic types of welding joints, types of base metal edge preparations, and types of welds. You will also be able to describe the AWS welding symbol and demonstrate how it is used on a welding drawing.

INSTRUCTIONS
Carefully read Headings 3.1 through 3.16 of the text. Also study Figures 3-1 through 3-30 in the text. Then, complete the following problems.

1. Name the five basic joints shown below.

   A. _____
   B. _____
   C. _____
   D. _____
   E. _____

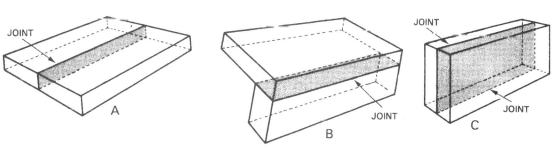

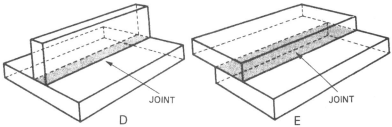

2. Label the areas and parts of the completed weld below.

   A. _____
   B. _____
   C. _____
   D. _____
   E. _____
   F. _____
   G. _____
   H. _____
   I. _____

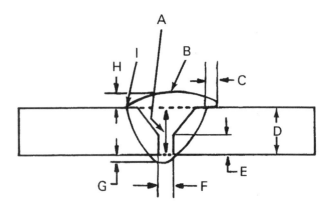

3. Name the various butt grooves shown below.

   A. _____
   B. _____
   C. _____
   D. _____
   E. _____
   F. _____
   G. _____
   H. _____
   I. _____
   J. _____
   K. _____
   L. _____

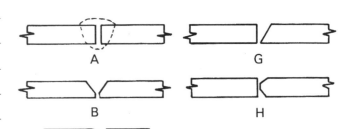
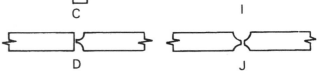
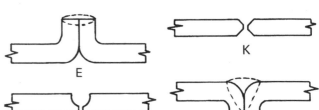

Name: _____ Lesson 3 Reading Welding Symbols 31

4. Name each type of corner joint shown below.

   A. _____
   B. _____
   C. _____
   D. _____
   E. _____
   F. _____
   G. _____
   H. _____
   I. _____
   J. _____

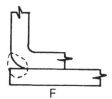
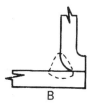
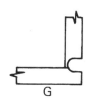
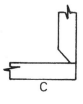
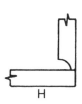
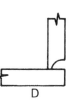
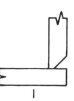

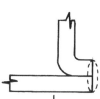

5. Identify each "T" joint or edge joint below.

   A. _____
   B. _____
   C. _____
   D. _____
   E. _____
   F. _____
   G. _____
   H. _____

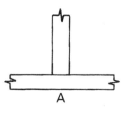
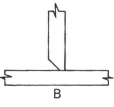
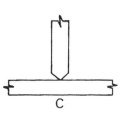

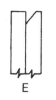

6. What is the difference between a welding symbol and a weld symbol?
   A. The weld symbol gives all the information required to make a weld; the welding symbol does not.
   B. The weld symbol is part of the welding symbol.
   C. The welding symbol is part of the weld symbol.
   D. The weld symbol tells how strong the weld is to be.
   E. There is no difference.

6. _____

7. Certain information may be given on a welding symbol. This information is always found in the same location on the welding symbol. Match the information given with the letters indicating the letter or symbol shown on the welding symbol below.

   Root opening

   Specifications, process, and other special notes

   Groove angle

   Weld finish required

   Length of the weld

   Pitch of the weld

   Weld-all-around symbol

   Field weld symbol

   Contour symbol

7. A. _____
   B. _____
   C. _____
   D. _____
   E. _____
   F. _____
   G. _____
   H. _____
   I. _____

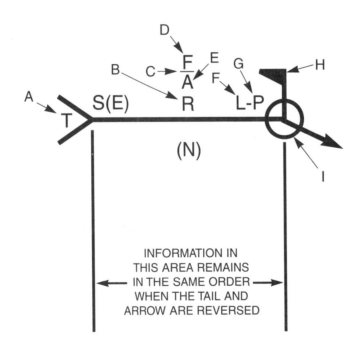

8. *True or False?* A weld is to be made on the side of the joint that the arrow touches. The symbol for that weld is always shown below the reference line.

8. _____

9. *True or False?* The vertical line used in the bevel, fillet, and J-groove joint symbol is always drawn to the left side of the symbol drawing.

9. _____

10. Given the information in the drawings below, first add the dimensions to the drawings in A through D. Then sketch in the completed welds on the partial pieces to the right of the illustration below.

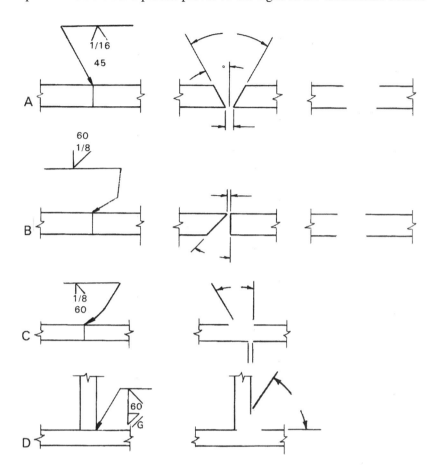

11. For the following welding symbols, draw the two pieces of metal as they would be prepared prior to welding and dimension the angles of spaces required.

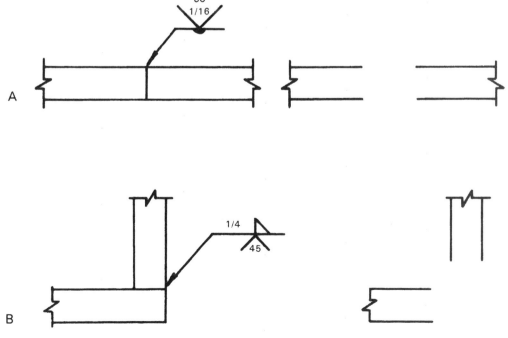

12. Show the shape of the weld face for the fillet welds made from these welding symbols.

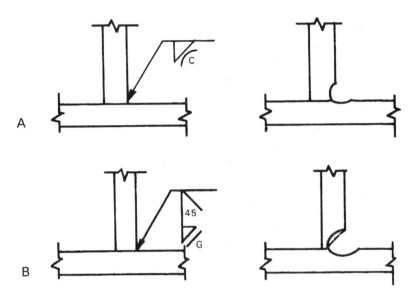

13. For the weld finish symbols given below, what process is used to finish the surface of the finished weld?

    C _____

    G _____

    M _____

14. The size of the weld and the effective throat size is always given to the _____ end of the _____ symbol.

    14. _____
        _____

15. The size of a groove weld is the depth to which the base metal is _____. The "S" size is the depth to which the metal is prepared.

    15. _____

16. A weld that is made in repeated short lengths is called a(n) _____ weld.

    16. _____

17. Draw and dimension the intermittent weld indicated by the following symbol.

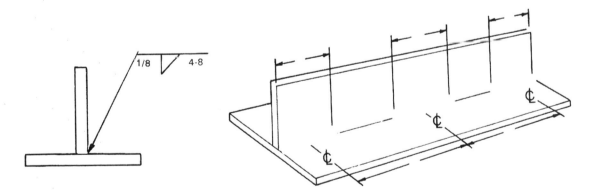

18. Draw and dimension the finished weld indicated by the welding symbol below.

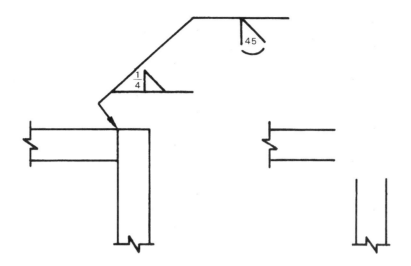

19. Draw the sides of the slot, the completed weld, and dimension the slot and weld that is required for the welding symbol below.

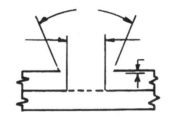

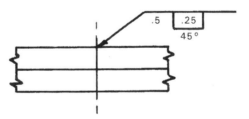

20. Answer the following questions about the weld symbols shown below.

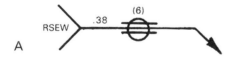

A. How wide is the weld in "A"?

B. What kind of a weld symbol is shown in "A"?

C. What welding process is used in "B"?

D. How many welds are made in "B"?

E. On what side of the metal is the weld in "B" made?

20. A. _____

B. _____

C. _____

D. _____

E. _____

Lesson 4A

WELDING AND CUTTING PROCESSES

Name _____ Date _____ Lesson Grade _____
Class _____ Instructor _____

LEARNING OBJECTIVE
★ You will be able to identify and describe eight welding and cutting processes.

INSTRUCTIONS
Carefully read Headings 4.1 through 4.8 of the text. Also study Figures 4-1 through 4-13 in the text. Then, complete the following problems.

1. Match the following welding or cutting processes listed on the left with their accepted American Welding Society (AWS) abbreviations.

    TS   A. Torch brazing   1. A. _____
    FCAW   B. Flux cored arc welding   B. _____
    TB
    OAW   C. Oxyacetylene welding   C. _____

                    D. Torch soldering   D. _____

2. Name the parts indicated below.

    A. _____
    B. _____
    C. _____
    D. _____
    E. _____
    F. _____
    G. _____
    H. _____

3. Give the temperatures of each item listed below in both °F and °C.

    A. A neutral oxyacetylene flame: _____°F (_____°C)

    B. Soldering is done below: _____°F (_____°C)

    C. Brazing is done above: _____°F (_____°C)

    D. An oxidizing oxy-LP gas flame: _____°F (_____°C)

4. Gas tungsten arc welding (GTAW) is particularly useful for welding _____ _____, aluminum, titanium, and many nonferrous metals.

4. _____

5. When GTAW, the welding amperage is determined by the _____.
   A. metal thickness
   B. electrode diameter
   C. type of metal welded
   D. All of the above.

5. _____

6. Match the welding or cutting processes listed on the right with the accepted AWS abbreviations listed on the left.

   GMAW   A. Gas tungsten arc welding
   OFC
   GTAW   B. Gas metal arc welding
   SMAW   C. Shielded metal arc welding
                 D. Oxyfuel gas cutting

   6. A. _____
   B. _____
   C. _____
   D. _____

7. Name the items involved in SMAW indicated below.

   A. _____
   B. _____
   C. _____
   D. _____
   E. _____
   F. _____

8. Explain why oxyfuel gas cutting is sometimes called "burning." _____

_____

_____

9. What is the major difference between the parts of an OAW and OFC outfit? _____

_____

10. List three pieces of safety equipment that must be worn when oxyacetylene welding or cutting. _____

_____

_____

_____

11. What is the correct and full name of MAPP gas? _____

_____

12. A _____ is used to clean the metal when TB.

12. _____

13. Why does TB cause less metal warping than is caused by OAW? _____

_____

Name: _____ Lesson 4A Welding and Cutting Processes 39

14. Possible health hazards may be involved when brazing metals containing _____.
   A. zinc
   B. beryllium
   C. phosphorous
   D. cadmium
   E. All of the above.

14. _____

15. Name the parts indicated on the gas tungsten arc welding station illustrated below.

   A. _____
   B. _____
   C. _____
   D. _____
   E. _____
   F. _____

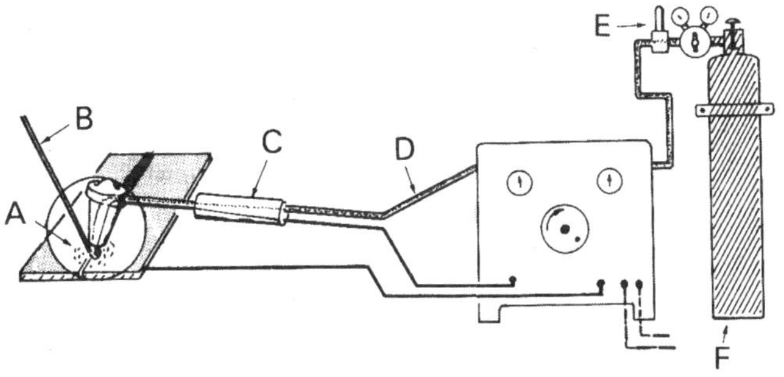

16. List three pieces of safety equipment that must be worn when SMAW, arc welding, or cutting. _____

_____

_____

_____

17. GTAW torches may be _____ cooled or _____ cooled.

17. _____

18. Name the indicated parts on the GMAW outfit illustrated below.

   A. _____
   B. _____
   C. _____
   D. _____
   E. _____
   F. _____
   G. _____
   H. _____
   I. _____
   J. _____
   K. _____
   L. _____

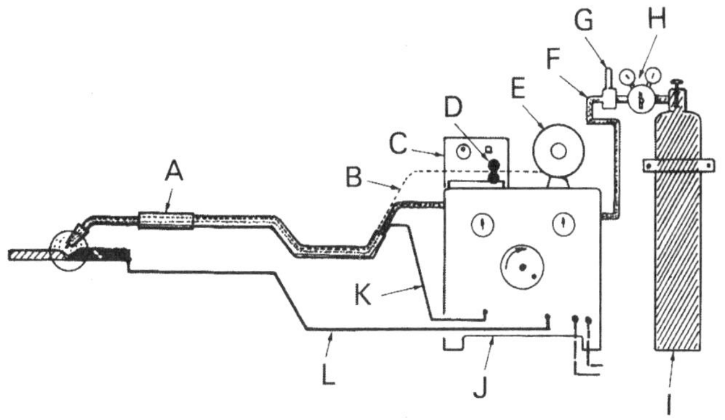

19. In FCAW, where is the flux material stored so that it is in constant use when welding? _____

_____

20. When welding with FCAW and GMAW, the current is changed by changing the _____ _____ feed speed.

20. _____

Copyright Goodheart-Willcox Co., Inc.

Lesson 4B

WELDING AND CUTTING PROCESSES

Name _____ Date _____ Lesson Grade _____
Class _____ Instructor _____

LEARNING OBJECTIVE
★ You will be able to identify and describe eight additional welding and cutting processes.

INSTRUCTIONS
Carefully read Headings 4.9 through 4.16 of the text. Also study Figures 4-14 through 4-21 in the text. Then, complete the following problems.

1. Match the following processes with their approved AWS abbreviations.

   AOC   A. Shielded metal arc cutting   1. A. _____
   SMAC
   FW   B. Plasma arc cutting   B. _____
   PAC   C. Oxygen arc cutting   C. _____
               D. Flash welding   D. _____

2. Match these processes with their approved AWS abbreviations.

   RSW   A. Gas tungsten arc cutting   2. A. _____
   RPW
   GTAC   B. Air carbon arc cutting   B. _____
   AAC   C. Resistance spot welding   C. _____
               D. Projection welding   D. _____

3. A _____ joint is used in resistance spot welding to weld together two or more pieces.

   3. _____

4. When resistance spot welding, the operator must wear _____ _____ and _____.

   4. _____

5. What two purposes does the shielding gas serve in GTAC?
   A. Supports combustion and helps to cut the metal.
   B. Melts the base metal and cools the electrode.
   C. Shields the metal from oxidation and melts the base metal.
   D. Shields the metal from oxidation and blows the oxidized metal away.
   E. None of the above.

   5. _____

Copyright Goodheart-Willcox Co., Inc.

6. When GTAC, an arc is struck between the tungsten _____ and the base metal.

6. _____

7. *True or False?* The air used in the air carbon arc cutting process may be supplied from a compressed air cylinder.

7. _____

8. A hollow electrode is used in the _____ process.
   A. GTAC
   B. AAC
   C. AOC
   D. SMAC
   E. Both A & C.

8. _____

9. The electrode used when SMAC is heavily covered with _____.

9. _____

10. Name the parts indicated in the illustration below.
    A. _____
    B. _____
    C. _____
    D. _____
    E. _____
    F. _____

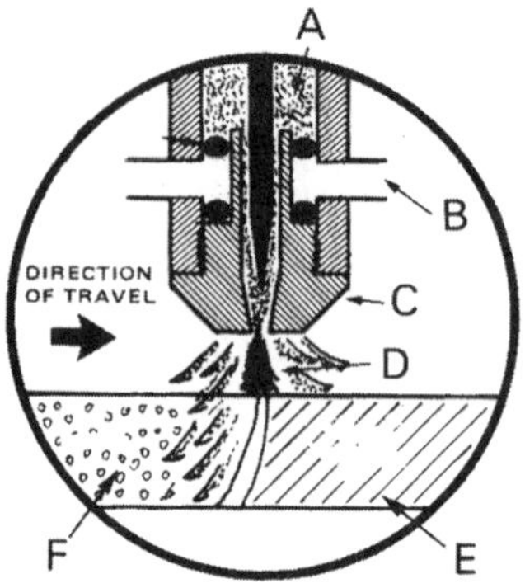

11. The PAC process cuts _____, stainless steel, and most other metals rapidly.

11. _____

12. What causes the base metal to be blown away from the heated area when PAC? _____

13. Describe what is meant by "projections" in the projection welding process. _____

14. Name the parts indicated on the RPW schematic illustrated below.
    A. _____
    B. _____
    C. _____
    D. _____
    E. _____
    F. _____
    G. _____
    H. _____

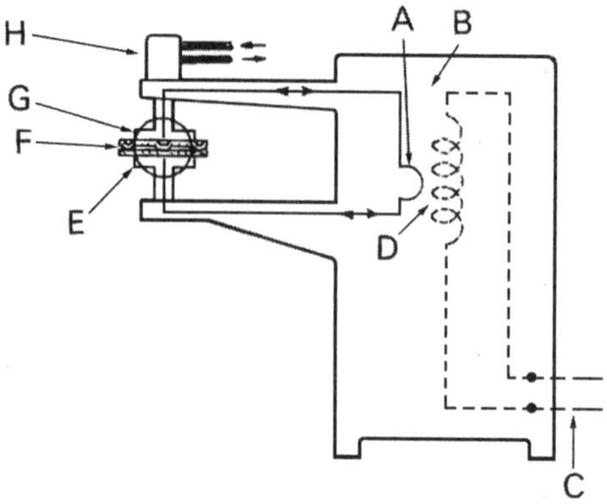

15. Name the areas of the completed friction weld shown below.

    A. _____

    B. _____

    C. _____

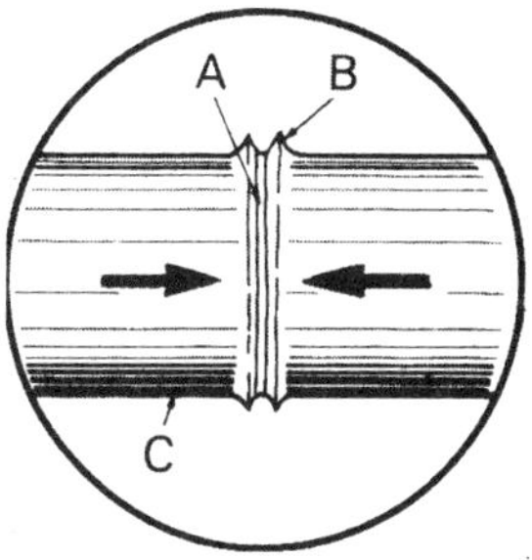

16. What causes the ends of the two pieces of metal to melt when doing FW? _____
    _____

17. Explain how the pieces are caused to weld together after they are heated to the molten state in FW. _____
    _____
    _____

18. List three pieces of safety equipment that must be worn during all cutting operations. _____
    _____
    _____
    _____

19. Oxygen from the oxygen cylinder is used for _____ when AOC.   19. _____
    A. cutting the heated metal
    B. cooling the heated metal
    C. cooling the electrode
    D. cooling the welder's hand
    E. cooling the electrode holder

20. Name the parts indicated in the AAC drawing below.

    A. _____

    B. _____

    C. _____

    D. _____

    E. _____

    F. _____

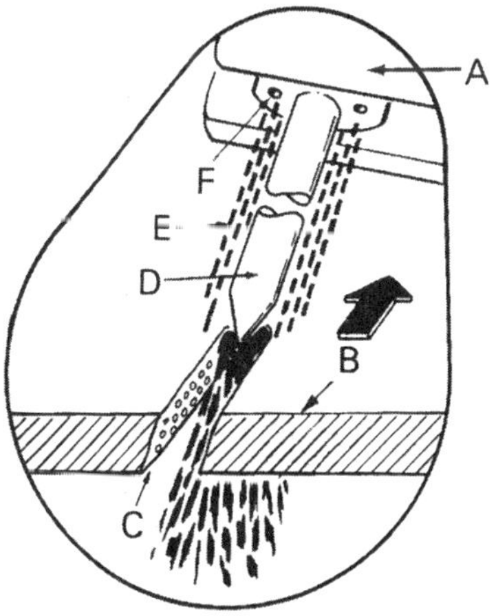

Lesson 4C

WELDING AND CUTTING PROCESSES

Name _____ Date _____ Lesson Grade _____
Class _____ Instructor _____

LEARNING OBJECTIVE
* You will be able to describe welding and cutting processes that are often used in industry.

INSTRUCTIONS
Carefully read Headings 4.17 through 4.34 and Figures 4-32 through 4-39 of the text. Then, complete the following problems.

1. Using Figure 4-38 on page 89 of the text, give the AWS abbreviations for each of the welding processes listed below.

    A. Chemical flux cutting   1. A. _____

    B. Carbon arc cutting   B. _____

    C. Infrared brazing   C. _____

    D. Electrogas welding   D. _____

    E. Arc stud welding   E. _____

2. Submerged arc welding is called "submerged" because the   2. _____
   welding is done under _____.
    A. water
    B. a heavy gas shield
    C. heavy metal
    D. a layer of molten metal
    E. a heavy layer of flux

3. The molten metal is kept from running out of the weld in ESW as   3. _____
   the weld moves up by sliding _____ _____.

4. Which welding process is generally used to attach bolts, screws, rivets, and spikes to metal surfaces? _____

_____

Copyright Goodheart-Willcox Co., Inc.

45

5. Name the parts of the CW process indicated below.

   A. _____
   B. _____
   C. _____
   D. _____

6. Which welding process uses a tremendous shock wave to fuse (weld) the pieces together? _____

7. In FOW, what is used to heat the parts to be welded? _____

8. In FRW, how is enough friction heat created to cause steel to melt and weld? _____

9. When ultrasonic welding, how is the degree of heating controlled? _____

10. When welding with a ruby laser beam, the energy builds up until it escapes from the weaker or poorer of the _____.

    10. _____

11. *True or False?* The electron beam welder produces welds on thick metal that are deep and wide.

    11. _____

12. Plastic is generally welded at temperatures of _____ °F to _____ °F by heated air or gas.

    12. _____

13. *True or False?* When lance cutting, the oxygen lance used is burned up in the cutting process.

    13. _____

14. When cutting under water, the oxyfuel cutting flame is kept from going out by the _____.
    A. pressure of the cutting oxygen
    B. seawater
    C. metal air jacket around the flame
    D. movement of the torches

    14. _____

15. Flux powder reduces the formation of _____ _____ and makes cutting easier when FOC.

    15. _____

16. The color of steel as it is heated indicates the temperature of the metal fairly accurately. Give the following temperatures:

    Bright yellow: _____°F (_____°C)

    Orange red: _____°F (_____°C)

    Deep straw: _____°F (_____°C)

17. Using Figure 4-39 on page 90 of the text, give the welding or cutting process indicated by the letter abbreviations below.

    A. EBC: _____

    B. PSP: _____

    C. LBW: _____

    D. GMAW: _____

18. Electron beam welding uses the energy of a focused stream of _____ to heat and weld metals.

    18. _____

19. *True or False?* Laser welding beams can be reflected in any direction by mirrors.

    19. _____

20. Even when cold welding, the welder should wear _____
    _____
    _____.

Lesson 5A

OXYFUEL GAS WELDING EQUIPMENT AND SUPPLIES

Name _____ Date _____ Lesson Grade _____
Class _____ Instructor _____

LEARNING OBJECTIVE
★ You will be able to determine what parts and equipment are assembled to make an oxyfuel gas welding station. You will be able to describe how the parts are constructed and assembled. You will also be able to list the supplies needed for oxyfuel gas welding.

INSTRUCTIONS
Carefully read Headings 5.1 through 5.10 of the text. Also study Figures 5-1 through 5-65 in the text. Then complete the following problems.

1. Label the parts of the oxyacetylene welding outfit shown below.

   A. _____
   B. _____
   C. _____
   D. _____
   E. _____
   F. _____
   G. _____
   H. _____
   I. _____
   J. _____
   K. _____
   L. _____
   M. _____
   N. _____
   O. _____
   P. _____

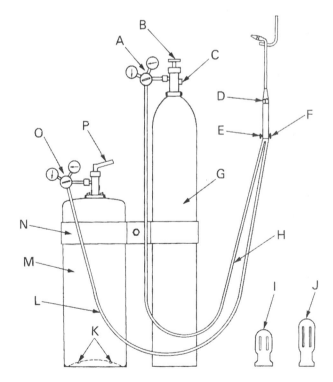

Copyright Goodheart-Willcox Co., Inc.

49

2. Liquid oxygen is generally stored and shipped in a _____ flask.

2. _____

3. Oxygen cylinder construction specifications are prepared by the _____. No part of the cylinder is less than _____ inch thick. The cylinder must withstand _____ psig of hydrostatic pressure.

3. _____
_____
_____

4. Prior to putting C_2H_2 gas in them, acetylene cylinders are filled with two or three materials. Name two of these substances. _____
_____

5. The rate at which acetylene can be drawn off or removed from an acetylene cylinder is dependent upon the _____.
   A. amount of acetylene left in the cylinder
   B. temperature of the cylinder
   C. number of cylinders hooked up to the system
   D. All of the above.
   E. None of the above.

5. _____

6. Mixing _____ and water will produce acetylene gas.

6. _____

7. Oxygen cylinders may contain 2000-2200 psig pressure when full. Welding torches operate at _____ to _____ psig. Therefore, a pressure regulator is required to reduce and control the pressure.
   A. 3 to 50
   B. 1 to 30
   C. 0 to 30
   D. 0 to 50
   E. 1 to 50

7. _____

8. Label the parts of the single-stage nozzle type regulator schematic shown below.

A. _____
B. _____
C. _____
D. _____
E. _____
F. _____
G. _____
H. _____
I. _____
J. _____
K. _____
L. _____
M. _____

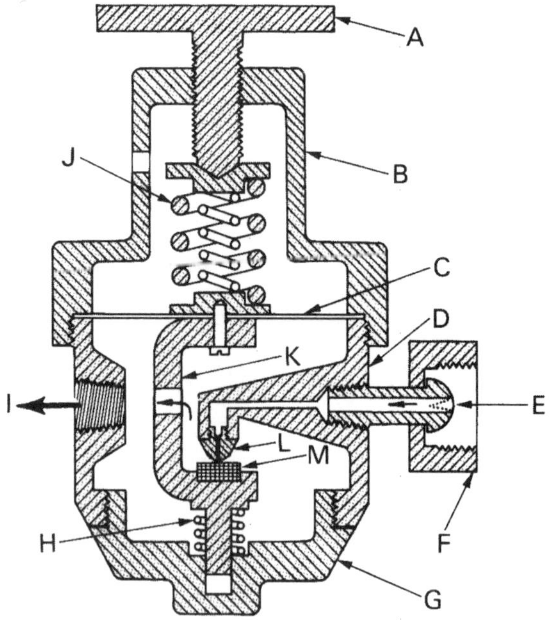

9. A two-stage regulator reduces the cylinder pressure to a working pressure in two stages. The lowest first stage pressure is generally _____ kPa (metric) of pressure.
   A. 200
   B. 34.5
   C. 5
   D. 1379
   E. 2000

9. _____

10. A regulator is turned completely off when the regulator screw is turned all the way _____.

10. _____

11. Why is the area above 15 psig marked with a red band on some acetylene regulators? _____

12. Hoses are made in three colors which are _____, _____, and _____. _____ is usually used for the fuel gas.

12. _____

13. Hose nuts are not interchangeable. The acetylene or fuel gas nut is different from the oxygen nut. The acetylene nut has a _____ cut around it, left hand threads, and it may have _____ stamped on it.

13. _____

14. Label the parts of the welding torch shown below.
   A. _____
   B. _____
   C. _____
   D. _____

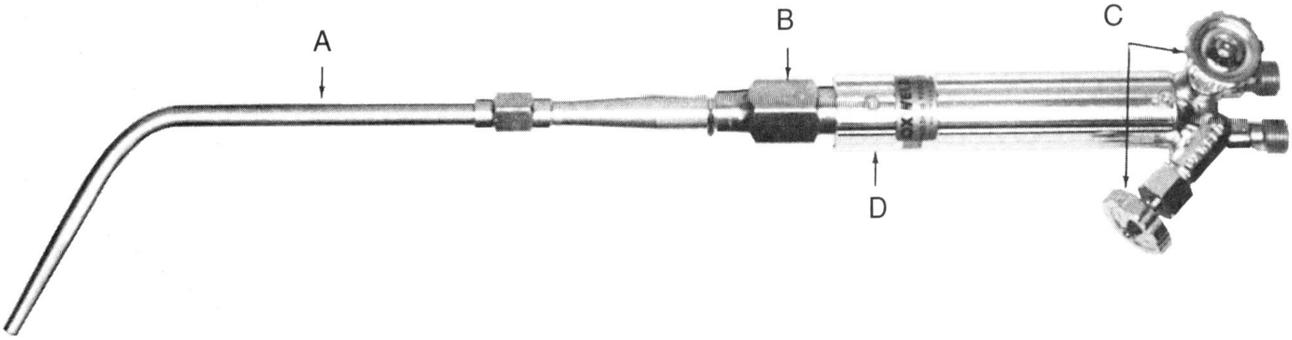

15. Broaches may be used to clean the welding tip orifice. Some welders prefer to use the correct size _____ drill to clean the tip orifice.

15. _____

16. Give the recommended welding lens shade number for use in the welding or cutting processes listed.

   A. Medium gas welding   16. A. _____

   B. Heavy gas welding   B. _____

   C. Light gas welding   C. _____

17. Flint and steel _____ lighters are generally used to light the   17. _____
    oxyfuel gas welding or cutting flame.

18. List six gases used with the oxyfuel gas process. _____

    _____

    _____

    _____

    _____

    _____

19. MAPP gas, also called methylacetylene propadiene, is safer than   19. _____
    acetylene and can be used under water at pressures over _____
    psig or _____ kPa.   _____

20. How do flashback arrestors prevent flashbacks from occurring? _____

    _____

    _____

Lesson 5B

OXYFUEL GAS WELDING AND CUTTING SAFETY

Name _____ Date _____ Lesson Grade _____

Class _____ Instructor _____

LEARNING OBJECTIVE

★ You will be able to safely use oxyfuel gas welding and cutting equipment.

INSTRUCTIONS

Carefully read Headings 5.2.1, 5.2.3, 5.3.1, 5.3.2, 5.3.4, 5.4.1, 5.4.6, 5.8, 5.10.2, 5.10.7, 6.5.1, 6.5.2, 6.5.5, 6.5.7, 6.12.3, 6.14, 6.14.1, 6.14.3, 8.5.1, 8.10, 9.15, and 10.1 of the text. Also study the related Figures in the text. Then complete the following problems.

1. The oxygen cylinder valve is constructed using a back seating valve to seal the stem from leakage. Therefore, when opening the oxygen cylinder, the valve must be _____.
   A. fully closed
   B. half way open
   C. fully opened
   D. whatever is required for a neutral flame
   E. at the correct pressure position

   1. _____

2. *True or False?* Cylinders must always be kept valve end up. When the cylinder is not in use, the cylinder valve must be closed whether the cylinder is full or empty.

   2. _____

3. Liquid oxygen is stored at _____°F (_____°C). It may cause freeze burns on the eyes or skin if it comes in contact with them.

   3. _____

4. Name seven organic materials that must be kept away from oxygen to prevent fires from occurring.

   _____
   _____
   _____
   _____
   _____
   _____

Copyright Goodheart-Willcox Co., Inc.

53

5. If clothing has been saturated by oxygen, it becomes highly flammable. It should be removed and not worn again for at least ____ minutes or until no oxygen remains in it.

6. *True or False?* Open flames and smoking are not permitted in any area where acetylene or oxygen is stored, used, or handled.

7. *True or False?* The storage of acetylene (C_2H_2) in its gaseous form is not permitted at pressures above 15 psig (103.4 kPa).

8. Fuse plugs are used to permit acetylene to be released from the acetylene cylinder in case of a fire. This will prevent the cylinder from exploding if overheated. The fuse plugs will melt at ____°F (____°C).

9. *True or False?* A lighted torch should never be aimed toward the cylinders or hoses.

10. Piping made of ____ must never be used in the presence of acetylene.

11. Which statement is *not* true with respect to acetylene cylinders and their use?
    A. Acetylene smells like garlic.
    B. Too much acetylene in the air can cause dizziness.
    C. Always leave the cylinder valve wrench, wheel, or key in place for emergency shutoff.
    D. Acetylene is stored at 2000 psig (13790 kPa) pressure.
    E. Flames should be kept away from the cylinder fuse plugs.

12. The oxygen and acetylene cylinder valves are closed when they are turned all the way ____.

13. Oxygen and acetylene regulators are closed when the regulator adjusting crew is turned all the way ____.

14. The filter lenses worn when oxyfuel gas cutting or welding protect the eyes from ____ and ____ rays.

15. *True or False?* Fuel cylinder valve threads are usually left-hand threads. Oxygen cylinder valves have right-hand threads. The thread diameters are also different. This is done to prevent connecting the wrong regulator to a cylinder.

16. Cylinder valves are protected from damage, while they are stored or moved, by threading on a properly fitting ____ ____.

17. Cylinders in use or in storage should be stored vertically. They should be held upright by a safety chain or metal band. They should be moved using a ____ truck.

18. A flashback is an extremely dangerous occurrence. If a flashback occurs, first close the torch ____ valve and then close the ____ torch valve. The regulator adjusting screws should then be turned out and the cylinders closed.

5. _____

6. _____

7. _____

8. _____

9. _____

10. _____

11. _____

12. _____

13. _____

14. _____
    _____

15. _____

16. _____

17. _____

18. _____
    _____

Name: _____   Lesson 5B Oxyfuel Gas Welding and Cutting Safety

19. Why shouldn't you place the welding rod against your body when changing the hand position on the rod? _____

20. List the four types of clothing the welder should wear when welding overhead. _____

21. List three metals whose fumes are toxic. These metals must be welded, cut, soldered, or brazed only when the ventilation is excellent. _____

22. The flame should always be _____ when the torch is not in your hand.

22. _____

23. Why shouldn't you stand in front of the regulator gauges when the cylinder is being turned on? _____

24. Tanks and containers that hold flammable or explosive materials should be welded only under the supervision of a qualified _____ engineer.

24. _____

25. The torch valves on the welding torch body must be turned _____ before the cutting attachment is connected or disconnected.

25. _____

Lesson 6A

OXYFUEL GAS WELDING—TURNING THE OUTFIT ON AND OFF

Name _____ Date _____ Lesson Grade _____
Class _____ Instructor _____

LEARNING OBJECTIVE
* You will be able to turn on, light and adjust the flame, and turn off an oxyacetylene welding outfit equipped with a positive-pressure type torch.

INSTRUCTIONS
Carefully read Headings 5.2.1, 5.5, 5.9, 6.5.1, 6.5.4, 6.5.5, 6.5.7, 6.5.8, and 6.14.3 of the text. Also study the related figures in the text. Then complete the following problems.

1. The pressure in a full oxygen cylinder is about _____ psig (_____ kPa).
   1. _____

2. Hoses should never be interchanged (switched). If _____ is passed through a hose previously used for acetylene, a combustible mixture might form and cause a flame in the hose.
   2. _____

3. Why is it recommended that you do not carry matches and other combustible items in your pockets? _____
   _____
   _____
   _____

4. *True or False?* The torch should not be pointed at another person while it is being lighted.
   4. _____

5. If the low-pressure gauge reading continues to creep upward after the torch valve is turned off, this indicates a leaking regulator nozzle and seat. If this happens, turn off the _____ valve immediately, or the low-pressure gauge may burst.
   A. regulator
   B. torch
   C. cylinder
   D. gauge
   E. station
   5. _____

Copyright Goodheart-Willcox Co., Inc.

6. Which of the flames shown below is neutral? _____

Which of the flames shown below is oxidizing? _____

Which of the flames shown below is carburizing? _____

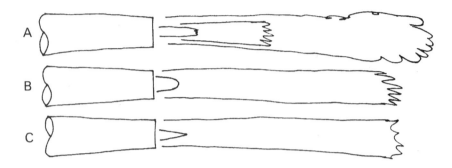

7. The eight steps required to open an oxyacetylene welding or cutting outfit are given below and on the next page. They are not in the correct order. Place them in order by writing the correct letter in the appropriate blank.

   A. Open the cylinder valve very slowly (counterclockwise) to prevent damage to the regulator diaphragm from the pressure and heat of 2000 psi (13,790 kPa). When the regulator high-pressure gauge reaches its highest reading, turn the cylinder valve all the way open. This is necessary because the oxygen cylinder valve has a double seat or a back-seating valve. In the full out position, this seat closes any possible opening along the valve stem through which the high-pressure oxygen could escape.

   B. Adjust the oxygen torch pressure. Open the torch oxygen valve one turn. Turn the oxygen regulator adjusting screw in (clockwise) until the low-pressure oxygen gauge indicates the pressure which is correct for the tip orifice. Then turn off the oxygen torch valve. The regulator pressures have now been adjusted to approximately the proper levels. **Important**—Check the low-pressure gauge for an indication of possible regulator defects. To prevent damage to the needle valves and seats, use only fingertip force to open and close the torch valves.

   C. Visually inspect the equipment for condition.

   D. Slowly open the acetylene cylinder valve 1/4 to 1/2 turn counterclockwise. This will usually provide adequate flow and will permit rapid closure in an emergency. Use the proper size wrench, and leave it on the valve in case an emergency shutoff is needed.

   E. Turn the regulator adjusting screws all the way out (counter-clockwise) before opening the cylinder valves. This prevents damage to the regulator diaphragm.

   F. Open the acetylene torch valve one turn. Turn the acetylene regulator adjusting screw slowly in (clockwise) until the low-pressure acetylene gauge indicates a pressure that is correct
   (Continued)

1. _____
2. _____
3. _____
4. _____
5. _____
6. _____
7. _____
8. _____

for the tip size. (Never use acetylene gas at a gauge pressure above 15 psig (103.4 kPa). An approximate setting may be arrived at, as shown in the table, Figure 6-8 in the text. Turn off the acetylene torch valve using finger-tip force only. **Important**—Check the low-pressure gauge for an indication of possible regulator defects. See Heading 6.5.5 in the text.

G. Stand to one side of the regulator when opening the cylinder valves. A burst regulator or gauge could cause severe injury.

H. Inspect regulators.

8. The steps for lighting and adjusting the oxyacetylene torch to a neutral flame are given below and on the next page. Place them in order by writing the correct letter in the appropriate blank.

    A. After the acetylene is regulated, slowly open the oxygen valve on the torch. As the oxygen is fed into the flame, the brilliant acetylene flame turns purple and a small inner cone starts to form. This inner cone is light green in color. When first formed, the extremity of this inner cone will have a blurred and irregular contour. As you continue to turn on oxygen, the inner cone loses its blurred edge and becomes a round, smooth cone. Stop the adjustment at this point. Any increase in oxygen will result in an "oxidizing flame." (Too much oxygen will burn or oxidize the metal being heated.) The tip of this inner cone is the hottest part of the flame. See Figure 4-2 in the text for a color illustration. The correct quantities of gases for the smaller tip sizes may also be detected by listening to the torch flame. It should emit a soft purr, not a sharp irritating hiss, when correctly adjusted.

    B. If the torch burns with an irregular contour (feather) to the cone, the flame is called a "carburizing flame." There is an excess of acetylene. See Figure 4-2 in the text.

    C. Purge the system before lighting the torch. To ensure that the proper gases are in the respective hoses (no air or oxygen in the acetylene hose and no fuel gas in the oxygen hose), the system must be purged. This is done by allowing the acetylene to flow through the acetylene hose and oxygen to flow through the oxygen hose for a short time before lighting. **Note:** If Steps 1 through 8 in questions 7 through 14 are followed, the system will have been purged and the torch will be ready for lighting.

    D. After purging, crack the acetylene torch valve no more than 1/16 of a turn. Using a flint lighter, ignite the acetylene gas coming out of the tip.

    E. Continue to turn on the acetylene torch valve slowly until the acetylene flame jumps away from the end of the tip slightly. This indicates that the proper amount of acetylene is being fed to the tip. A quick flip of the torch should make the flame leap away from the tip 1/16" (1.6mm) and come back again. If the flame will not move back to the tip, too much acetylene has been turned on. (If the tip is worn, it may be difficult to make

1. _____

2. _____

3. _____

4. _____

5. _____

(Continued)

the flame jump away from the tip.) Another method for determining the correct amount of acetylene is to increase the flow until the flame becomes turbulent (rough) a distance of 3/4″ to 1″ (19mm to 25.4mm) from the torch tip. With the right amount of acetylene, the flame will no longer smoke, or release soot. Look at Figure 6-9 in the text and compare the flames. Refer to Figure 4-2 in the text for a color illustration of the flames.

9. The steps for shutting down an oxyacetylene outfit are given below. Place them in order by writing the correct letter in the appropriate blank.

   A. Wait until both the high- and low-pressure gauges on both the acetylene and oxygen regulators read zero.

   B. Close the hand valves on the torch. (Close the acetylene valve first.) This extinguishes the flame and eliminates the soot.

   C. Open the hand valves on the torch.

   D. Close both hand valves on the torch (lightly) and hang up the torch.

   E. Turn the adjusting screws on both the acetylene and oxygen regulators all the way out.

   F. Close the cylinder valves (tightly).

   1. _____
   2. _____
   3. _____
   4. _____
   5. _____
   6. _____

10. The oxyfuel gas flame should be lighted using a _____.

    10. _____

Job 6A-1

LEARNING OBJECTIVE	Name _____
✴ In this job, you will learn to safely and properly turn on the oxyacetylene welding outfit equipped with a positive-pressure torch. You will also learn how to light and adjust the flame and turn the outfit off.	Date _____ Class _____ Instructor _____ Grade _____

TURNING ON THE OXYACETYLENE WELDING OUTFIT

1. Work with a group of two or three students on an assigned oxyacetylene welding outfit. **Note:** Assume that you will be using a positive-pressure welding torch. Assume also that you will be welding on 1/8″ (3.2mm) base metal when setting the oxygen and acetylene pressures and selecting the torch tip size.

2. Determine the manufacturer of the tip. What is the manufacturer's name? _____

3. Determine the correct tip orifice drill size to be used for a metal thickness of 1/8″ (3.2mm) steel from Figure 6-8 in the text.

    What is the size? # _____

4. Determine the correct acetylene and oxygen pressures from the table in Figure 6-8 of the text. What are the pressures here in lbs./in² gauge (psig)?

    Acetylene _____ psig

    Oxygen _____ psig

5. The first student will tell the other member(s) of the group what is to be done to turn on the station, starting from the first step.

6. All members must agree that this is the correct thing to do and is in the correct order. After they agree, the first student will do what is agreed upon by the group.

7. The first student will proceed through each step, waiting until everyone agrees. Stop when step 8 is reached.

8. Have your instructor approve your work.

    Instructor's initials: _____

LIGHTING AND ADJUSTING THE OXYACETYLENE FLAME

1. The first student will continue on to light the flame and adjust it to neutral.

2. Before doing any step, the first student must tell the group what he or she intends to do. They all must agree before the step is accomplished.

3. The first student will continue step-by-step until the flame is lighted and a neutral flame is adjusted.

4. Have your instructor approve your work.

    Instructor's initials: _____

SHUTTING DOWN THE OXYACETYLENE OUTFIT

1. The first student will proceed to shut down the outfit completely.
2. The first student will tell the group what he or she intends to do.
3. There must be agreement among the group before each step is done.
4. The first student will continue through the proper steps until the welding outfit is completely and correctly shut down.

Each student will go through the entire turning on, lighting and adjusting, and shutting down procedure in the same manner as the first student. **Caution:** Be careful not to trap pressure in the high-pressure side of the regulator and in the high-pressure gauge. This will indicate an incorrect shut-down procedure.

Instructor's initials: _____

Lesson 6B

OXYFUEL GAS WELDING—RUNNING A CONTINUOUS WELD POOL

Name _____ Date _____ Lesson Grade _____
Class _____ Instructor _____

> **LEARNING OBJECTIVE**
>
> ★ You will be able to select the correct torch tip and oxygen and acetylene gas pressures. You will be able to run a continuous weld pool on mild steel. You will also be able to make an outside corner joint without a welding rod.
>
> **INSTRUCTIONS**
>
> Carefully read Headings 6.5.3, 6.5.5, 6.6, 6.7, and 6.71 of the text. Also study the tables in Figures 4-5 and 4-6 in the text. Then complete the following problems.

1. What should you do if the low-pressure gauge needle continues to rise or creep upward after the torch valve is turned off? _____
_____

2. To weld a 1/16" piece of low carbon steel, with a positive-pressure type welding torch, you should use a # _____ drill size tip, _____ psig of oxygen, and _____ psig acetylene.

   2. _____

3. To weld a 3/32" thick piece of metal with a positive-pressure torch, you should use a # _____ drill size tip, _____ psig of oxygen, and _____ psig of acetylene.

   3. _____

4. To weld a 1/8" thick piece of metal with an injector-type torch, you should use a # _____ drill size tip, _____ psig of oxygen, and _____ psig of acetylene.

   4. _____

5. When oxyfuel gas welding, the typical angle of the torch tip from the base metal surface is _____.
   A. 15° – 30°
   B. 20° – 30°
   C. 25° – 45°
   D. 30° – 45°
   E. 60° – 70°

   5. _____

Copyright Goodheart-Willcox Co., Inc.

6. In the space provided, draw three different torch motions which can be used when welding or running a continuous weld pool.

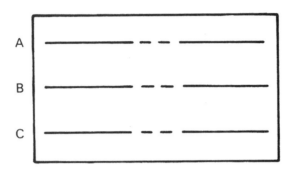

7. The tip of the inner flame should be _____ from the base metal surface.
   A. 1/32" – 1/8"
   B. 1/16" – 1/8"
   C. 1.6mm – 2.8mm
   D. 1.6mm – 3.5mm
   E. 1/8" – 1/4"

7. _____

8. List four things that an experienced welder can determine by watching the weld pool. _____
   _____
   _____
   _____

9. If the weld pool gets too wide, list two things that can be done with the torch tip to reduce the heat and decrease the pool size. _____
   _____

10. A well-made continuous weld pool will be straight, with even width, no holes, and good _____.

10. _____

Lesson 6B   Oxyfuel Gas Welding–Running a Continous Weld Pool   65

Job 6B-1

LEARNING OBJECTIVE
✶ In this job, you will learn how to run a continuous weld pool. This exercise teaches you how to control the molten metal without melting through to the other side.

Name _____
Date _____ Class _____
Instructor _____
Grade _____

1. Obtain a piece of 1/16″ (1.59mm) thick mild steel. It should be 3″ × 5″ (76mm × 127mm) in size.

2. Draw five evenly spaced lines on the metal surface using chalk or soapstone. These lines must run along the 5″ (127mm) length, one in the middle and two to each side. The outer lines should be 3/8″ (9.5mm) from the edges.

3. What type welding torch is used on your welding outfit?   3. _____
   A. Positive-pressure type
   B. Injector type

4. Complete the information below:

    #_____ drill size tip orifice

    _____ psig oxygen pressure

    _____ psig acetylene pressure

5. Place the metal between two firebricks, so that only the outer edges are supported by the bricks.

6. Turn on the outfit, light the flame, and adjust the flame to neutral.

7. If you are right-handed, start the weld pool at the right end of the metal. If you are left-handed, start at the left end of the metal.

8. The flame should be pointed in the direction of travel. The flame tip must be kept at the correct height above the base metal. An angle of 30°– 45° should be maintained between the torch tip and the base metal. This angle may vary up or down to heat or cool the weld pool if it is too narrow or too wide.

9. Complete one weld pool. Using your pliers to pick the metal up, inspect the bottom side for good, even penetration. The pool should be straight and have a uniform width.

10. Finish the other four weld pools. Use the lines drawn on the metal to keep the pool going in a straight line.

11. Inspect each weld pool as it is made, and attempt to correct any defect in your next pool.

12. If necessary, try again with a second piece of metal until five weld pools of good quality are complete on one piece.

Instructor's initials: _____

Job 6B-2

LEARNING OBJECTIVE

* In this job, you will make a square groove weld on an outside corner without the use of a welding rod

Name _____
Date _____ Class _____
Instructor _____
Grade _____

1. After completing the weld pool job, obtain two pieces of 1/16" (1.6mm) mild steel which measure 1 1/2" × 5" (38mm × 127mm).

2. Arrange the pieces as shown below to form an outside corner joint. The vertical piece should extend beyond the horizontal piece by 1/16" (1.6mm).

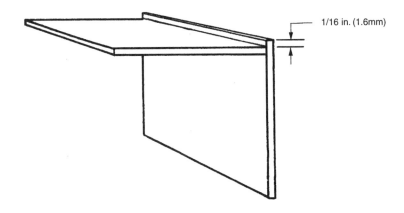

3. Tack weld the parts in three places, once at each end and once in the middle. This is done by melting down the overlap until it melts into the horizontal piece.

4. Start your weld from the end closest to your welding hand. Apply the torch to the overlapped edge and melt it down into the horizontal piece. A crescent-shaped or zigzag motion is suggested. Be certain to create a weld pool in the horizontal piece to permit the overlapped edge to melt (fuse) into it.

5. Make three of these outside corner joints without a welding rod.

INSPECTION
The weld bead should be slightly convex, with a uniform width, and with a small amount of penetration showing on the inside of the corner.

Instructor's initials _____

Lesson 6C

OXYFUEL GAS WELDING—WELDING MILD STEEL IN THE FLAT WELDING POSITION

Name _____ Date _____ Lesson Grade _____
Class _____ Instructor _____

LEARNING OBJECTIVE
✱ You will be able to select the correct welding rod. You will learn to select the correct tip size and welding gas pressure for a given metal thickness. You will also be able to weld several joints in the flat welding position.

INSTRUCTIONS
Carefully read Headings 5.10.3, 6.7.2 through 6.12, and 6.12.4 through 6.13 of the text. Also study Figures 6-18 through 6-34 in the text. Then complete the following problems.

1. Name five common metals used to make welding rods. _____
_____
_____
_____
_____
_____

2. Most welding rods are _____" long. They are packaged in _____ lb. (_____ kg) bundles.

2. _____

3. Welding rods made for oxyacetylene welding are the RG45, RG60, and RG65. The tensile strength of the RG60 welding rod is _____ psi or _____ Mpa.

3. _____

4. When welding, the welding rod should be held within _____" (_____ mm) of the flame tip and _____" (_____ mm) of the surface of the puddle to keep it preheated.

4. _____

5. The shape of the normal bead should be _____.
    A. concave
    B. convex
    C. flat

5. _____

6. The welding rod is added to the weld pool by _____.   6. _____
   A. allowing it to drip from the end of the molten rod
   B. touching it to the base metal ahead of the weld pool
   C. dipping the welding rod into the forward edge of the molten weld pool
   D. touching it to the base metal just behind the weld pool
   E. melting it off the end with the torch flame

7. List three reasons why a rod smaller than recommended should not be used. _____
   _____
   _____
   _____

8. The welding rod is only added to the lap and inside corner joint   8. _____
   when the weld pool runs forward at the outer edges to form a
   _____-shaped weld pool.
   A. T
   B. C
   C. Z
   D. O
   E. S

9. Define a weld made in the flat welding position. _____
   _____
   _____
   _____

10. Complete the drawing below of a V-groove butt weld in progress. Show the "keyhole" clearly in your work.

Job 6C-1

LEARNING OBJECTIVE

★ In this job, you will learn to produce a fillet weld on a lap joint in the flat welding position.

Name _____
Date _____ Class _____
Instructor _____
Grade _____

1. Obtain six pieces of 3/32" (2.4mm) mild steel. The pieces should measure 1 1/2" × 5" (38mm × 127mm).

2. For this metal, use the following:

    #_____ drill tip size

    _____" diameter welding rod

    _____ psig oxygen pressure

    _____ psig acetylene pressure

3. Make one of the following fillet welds after tack welding the joint in three places on both sides.

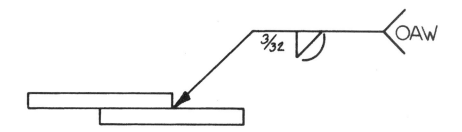

INSPECTION

1. Your weld bead should be even in width and slightly convex, with smooth and evenly spaced ripples in the beads.

2. After inspecting one sample, make two more lap welds on the remaining pieces.

3. Use your final weld for a test sample. Place the welded pieces into a vise. The weld should be just above the top of the vise jaws. **Caution:** The vise should have a safety screen around it. This will prevent the metal from flying across the room when it fails.

4. Bend one piece away from the other using a hammer. Continue to bend the piece back and forth until it fails. A strong weld will tear away from the tip or bottom piece. The weld should come away from the bottom piece with some metal from the surface. If it doesn't, the bottom surface was not melted enough before adding the welding rod, and the weld fusion is poor.

Instructor's initials: _____

Job 6C-2

LEARNING OBJECTIVE

* In this job, you will make a fillet weld on an inside corner joint in the flat welding position.

Name _____
Date _____ Class _____
Instructor _____
Grade _____

1. Obtain six pieces of 1/8″ × 1 1/2″ × 5″ (3.2mm × 38.1mm × 127mm) mild steel.

2. To weld this metal, use the following:

    #_____ drill tip size

    _____″ diameter welding rod

    _____ psig oxygen

    _____ psig acetylene

3. Make three of the following welds after tack welding the pieces three times on each side.

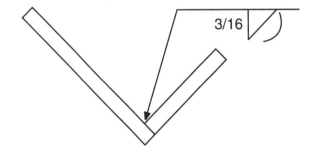

INSPECTION
All welds should be straight. The beads should be slightly convex and even in width, with smooth, evenly spaced ripples in the beads.

Instructor's initials: _____

Job 6C-3

LEARNING OBJECTIVE

★ You will learn to make a square-groove weld on an outside corner joint in the flat welding position.

Name _____
Date _____ Class _____
Instructor _____
Grade _____

1. Obtain six pieces of 3/32" (2.4mm) mild steel.
2. To weld this metal, use the following:

   #_____ drill tip size

   _____" diameter welding rod

   _____ psig oxygen

   _____ psig acetylene

3. Make three of the following welds after tack welding about 3" (76mm) apart on the weld side only.

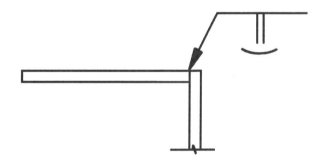

INSPECTION

All welds should be straight and slightly convex, with even width and smooth, even ripples in the bead. There should be little or no penetration showing on the other side of the weld.

Instructor's initials: _____

Job 6C-4

LEARNING OBJECTIVE

* You will learn to make a square-groove weld on a butt joint in the flat welding position.

Name _____
Date _____ Class _____
Instructor _____
Grade _____

1. Obtain six pieces of mild steel 1/16" (1.6mm) thick. The pieces should be 1 1/2" × 5" (38mm × 127mm).

2. For this weld, use the following:

    #_____ drill tip size

    _____" diameter welding rod

    _____ psig oxygen

    _____ psig acetylene

3. After tack welding about every 3" (76 mm) on the weld side, make three welds as shown below.

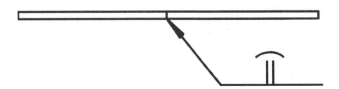

INSPECTION

All welds should have acceptable-looking beads as explained previously. Use your last joint as a test sample. Place it into a vise so that the middle of the weld bead is even with top of the vise jaws. **Caution:** Whenever a destructive test is made in a vise, the vise must have a screen or curtain around it. Bend the metal back and forth until it fails. Always make the bend toward the screen or curtain. A good weld will break in the base metal and not through the weld area.

Instructor's initials: _____

Job 6C-5

LEARNING OBJECTIVE

★ In this job, you will make a fillet weld on an inside corner and a V-groove weld on an outside corner in the flat welding position.

Name _____
Date _____ Class _____
Instructor _____
Grade _____

1. Obtain four pieces of 1/4″ × 1 1/2″ × 5″ (6.4mm × 38mm × 127mm) mild steel.

Prepare the edges as required, and perform two of the following welds:

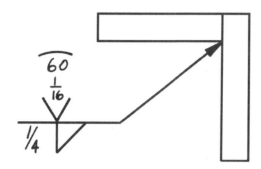

Caution: Before using the grinder, be certain you have passed the grinder safety test and are wearing grinding goggles.

2. For these welds, use the following:

   #_____ drill tip size

   _____″ diameter welding rod

   _____ psig oxygen

   _____ psig acetylene

Suggestion: Use 1/16″ (1.6mm) steel as spacers while you tack weld the pieces. If possible, remove the spacers prior to welding.

INSPECTION

The weld beads should be of the correct size, with straight beads, uniform width, and with smooth, even ripples in the beads.

Instructor's initials: _____

Job 6C-6

LEARNING OBJECTIVE

* In this job, you will make a fillet weld on a lap joint in the flat welding position, but on thicker metal.

Name _____
Date _____ Class _____
Instructor _____
Grade _____

1. Make two of the welds shown below using 1/4″ × 1 1/2″ × 5″ (6.4mm × 38mm × 127mm) mild steel.
2. Tack weld the joint in three places on each side.

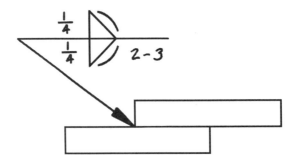

INSPECTION

Inspect visually. Each bend must be straight, with uniform width, and smooth, even ripples.

Instructor's initials: _____

Job 6C-7

LEARNING OBJECTIVE

★ In this job, you will learn to make a V-groove weld and a bevel-groove weld on a butt joint in the flat welding position.

Name _____
Date _____ Class _____
Instructor _____
Grade _____

1. Obtain four pieces of mild steel which measure 1/4″ × 1 1/2″ × 5″ (6.4mm × 38mm × 127mm). Make the welds shown below.

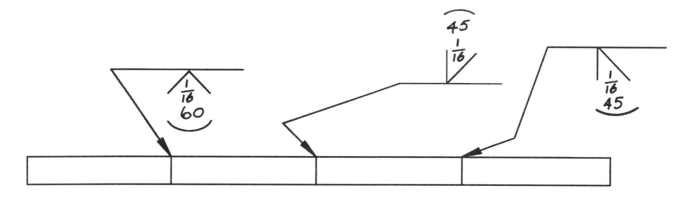

INSPECTION

Each butt weld must have full penetration throughout its length, an even bead width, a convex bead, and smooth, even ripples in the bead.

Instructor's initials: _____

Lesson 6D

OXYFUEL GAS WELDING—WELDING MILD STEEL IN THE HORIZONTAL WELDING POSITION

Name _____ Date _____ Lesson Grade _____
Class _____ Instructor _____

LEARNING OBJECTIVE
★ You will be able to weld the various types of joints in the horizontal welding position.

INSTRUCTIONS
Carefully read Headings 6.12.1 and 6.12.4 through 6.13 of the text. Also study Figures 6-34, 6-35, 6-40, 6-43, 6-46, and 6-47 in the text. Then complete the following problems.

1. through 8. Complete the drawings below to show the type of good or defective weld indicated.

   1. Properly made, with complete penetration and proper fusion at the toes of the weld.

   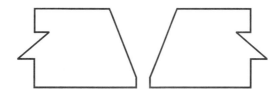

   2. Less than 100% penetration, with proper fusion and overlapping at the toes of the weld.

   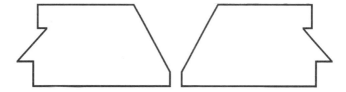

   3. Properly made, with good fusion at the toes of the weld and adequate penetration for a lap joint.

   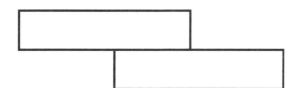

4. Incomplete fusion at the toe of the weld.

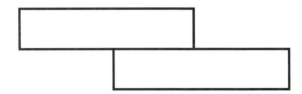

5. A fillet weld on a T-joint. This weld has undercutting on the vertical piece and has a smaller-than-normal weld size. Show the ideal weld size as a hidden (dashed) line.

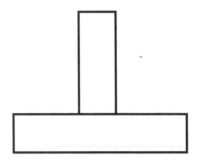

6. A butt weld, with undercutting at the toes and a weld face below the metal surface.

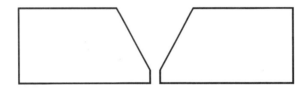

7. A fillet weld, with undercutting on the lower surface of the lower piece. Show a weld with a convex bead.

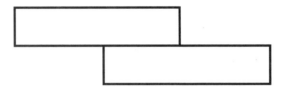

8. A weld with 100% penetration but with a convex bead.

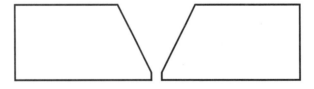

9. Define the horizontal welding position. _____
_____

10. To prevent the molten metal in the weld from sagging while making a horizontal weld, the torch tip and welding flame should point _____.
    A. straight along the weld line
    B. slightly downward from the weld axis (line)
    C. perpendicular to the base metal
    D. slightly upward from the weld axis (line)
    E. at the lower piece of metal

10. _____

Job 6D-1

LEARNING OBJECTIVE

★ In this job, you will learn to make a fillet weld on a lap joint in the horizontal welding position.

Name _____
Date _____ Class _____
Instructor _____
Grade _____

1. Obtain four pieces of mild steel which measure 3/32″ × 1 1/2″ × 5″ (2.4mm × 38mm × 127mm).

2. Use the following for these welds:

   #_____ drill tip size

   _____ diameter welding rod

   _____ psig oxygen

   _____ psig acetylene

3. Tack weld the pieces three times on each side of the metal while they are in the flat welding position to make the welds shown in step #5.

4. Place the metal in a positioning fixture or prop it against firebricks so the weld axis (line) is horizontal.

5. Make the following welds. The positioning of the weldment should be changed to permit all welds to be made in the horizontal welding position.

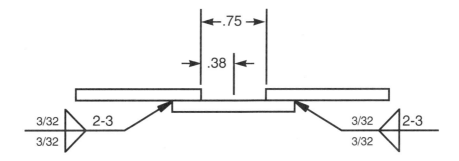

INSPECTION

Visually check each fillet weld. The beads should be straight, convex, have evenly spaced ripples, and good fusion between the bead and surface.

Instructor's initials: _____

Job 6D-2

LEARNING OBJECTIVE

★ In this job, you will learn to make a fillet weld on a T-joint in the horizontal welding position.

Name _____
Date _____ Class _____
Instructor _____
Grade _____

1. Obtain three pieces of 1/16" (1.6mm) mild steel measuring as follows:
   Two - 1 1/2" × 5" (38mm × 127mm)
   One - 3" × 5" (76.2mm × 127mm)

2. Use the following:

   #_____ drill tip size

   _____ diameter welding rod

   _____ psig oxygen

   _____ psig acetylene

3. Assemble the pieces to make the weld shown in step #5. Tack weld three places on each side of the weldment. This may be done with the metal in the flat position.

4. Place the weldment into the positioning fixture or prop it against firebricks. The weld axis must be horizontal and the weld face at 45°.

5. Make the welds shown below.

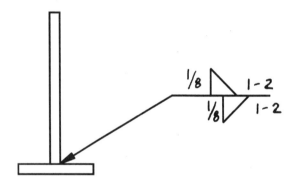

6. After completing step #5, make the following welds to complete the weldment.

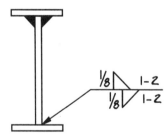

INSPECTION

Make a visual inspection as done previously. The dimensions of the staggered intermittent welds should be approximately as shown.

Instructor's initials: _____

Job 6D-3

LEARNING OBJECTIVE

★ In this job, you will learn to make a square-groove weld on a butt joint in the horizontal welding position.

Name _____
Date _____ Class _____
Instructor _____
Grade _____

1. Obtain four pieces of mild steel that measure as indicated below:
   Two 1/16" × 1 1/2" × 5" (1.6mm × 38mm × 127mm)
   Two 1/8" × 1 1/2" × 5" (3.2mm × 38mm × 127mm)

2. For the 1/16" (1.6mm) thick steel (Weld #1), use the following:

   #_____ drill tip size

   _____ diameter welding rod

   _____ psig oxygen

   _____ psig acetylene

3. For the 1/8" (3.2mm) thick steel (Weld #2), use the following:

   #_____ drill tip size

   _____ diameter welding rod

   _____ psig oxygen

   _____ psig acetylene

4. Tack weld the joints in three places to make the welds shown in steps #6 and #7. The tack welds may be made in the flat welding position.

5. Place the weldments into the positioning fixture prior to welding. The weld axis must be horizontal and the weld face vertical.

6. Weld #1:

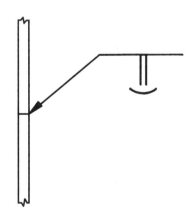

7. Weld #2:

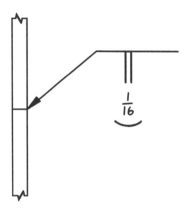

INSPECTION

Make a visual inspection as done previously. Be certain that the butt welds have 100% penetration. Bend each weld until it fails. The failure must occur in the base metal and not in the weld metal. This indicates a properly made butt weld. **Caution:** While performing the bend test, be certain to wear safety glasses. Also, bend only toward the curtain or screen around the vise.

Instructor's initials: _____

Lesson 6E

OXYFUEL GAS WELDING—WELDING MILD STEEL IN THE VERTICAL WELDING POSITION

Name _____ Date _____ Lesson Grade _____
Class _____ Instructor _____

LEARNING OBJECTIVE
★ You will be able to describe methods of welding thick metal sections. You will also be able to make welds in the vertical welding position.

INSTRUCTIONS
Carefully read Headings 6.12.2 and 6.12.4 through 6.12.6 of the text. Also study Figures 6-25 and 6-38 through 6-49 in the text. Then complete the following problems.

1. The _____ keeps the molten metal in the vertical weld from sagging and falling out of the weld pool.
   A. force of gravity
   B. force of the welding gas coming from the tip
   C. fact that the metal cools rapidly
   D. action of the hot welding rod
   E. air circulating around the weld

   1. _____

2. The torch tip should be held at a _____ angle to the base metal surface.
   A. 15° – 30°
   B. 30° – 60°
   C. 30° – 90°
   D. 15° – 60°
   E. 25° – 30°

   2. _____

3. Name the various areas and measurements of the V-groove weld below.

   A. _____
   B. _____
   C. _____
   D. _____
   E. _____

Copyright Goodheart-Willcox Co., Inc.

83

4. Name the nine butt joint welds shown below.

A. _____
B. _____
C. _____
D. _____
E. _____
F. _____
G. _____
H. _____
I. _____

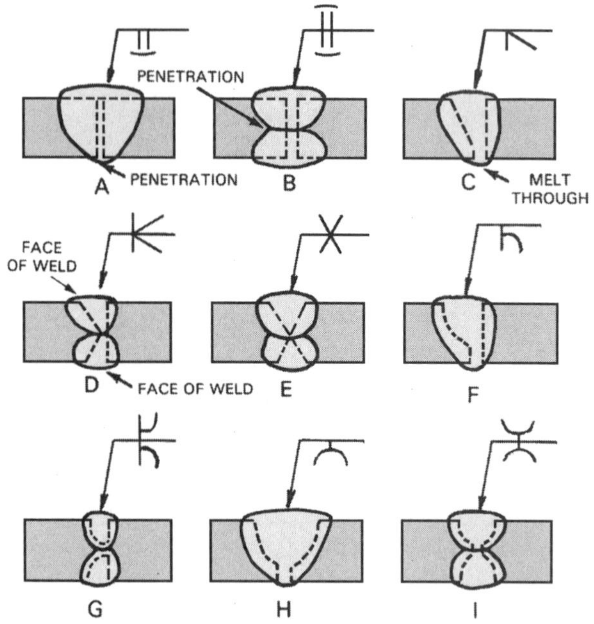

5. Directing the flame in a backhand manner tends to _____ the completed weld and relieves the welding _____ to a great extent.

5. _____
_____

6. Which of the following statements are true for multiple-pass or step-pass welds?
   A. They are generally on thick metal.
   B. They permit beads of a size that are easily handled.
   C. They permit welds thin enough to allow impurities to escape before the weld solidifies.
   D. All of the above.

6. _____

7. A "flip motion" tends to do what for the molten weld pool? _____
_____
_____

8. Describe or define "backhand welding." _____
_____
_____
_____

9. Define a "multiple pass weld." _____
_____
_____
_____

10. Define the term "vertical welding." _____
_____
_____
_____

Job 6E-1

LEARNING OBJECTIVE	Name _____
★ For this job you will make a fillet weld on a lap joint in the vertical welding position.	Date _____ Class _____ Instructor _____ Grade _____

1. Obtain three pieces of mild steel that measure 1/16" × 1 1/2" × 5" (1.6mm × 38mm × 127mm).

2. Use the following for these welds:

    #_____ drill tip size

    _____ diameter welding rod

    _____ psig oxygen

    _____ psig acetylene

3. Tack weld all the welded joints indicated in step #5 three times. This may be done in the flat welding position.

4. Place the weldment into a positioning fixture so the weld axis is vertical. You may prop the weldment against firebricks if a positioning fixture is not available.

5. Make the following four welds.

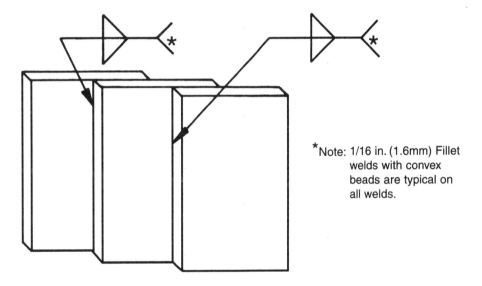

*Note: 1/16 in. (1.6mm) Fillet welds with convex beads are typical on all welds.

Note: 1/16" (1.6mm). Fillet welds with convex beads are typical on all welds.

6. The weldment may be repositioned as necessary to make all the welds in the vertical welding position and from the bottom up.

INSPECTION

Each fillet weld should be straight, convex, and with evenly spaced ripples. Good fusion between the bead and the surface should be indicated.

Instructor's initials: _____

Job 6E-2

LEARNING OBJECTIVE

★ For this job, you will weld a fillet weld on a T-joint in the vertical welding position.

Name _____
Date _____ Class _____
Instructor _____
Grade _____

1. Obtain three pieces of 1/8" × 1 1/2" × 5" (3.2mm × 38mm × 127mm) mild steel.
2. Use the following for these welds:

    #_____ drill tip size

    _____ diameter welding rod

    _____ psig oxygen

    _____ psig acetylene

3. Tack weld the weldment as shown in step #5. Tack welding may be done in the flat welding position.
4. Place the weldment in a welding fixture or prop it against several firebricks. The weld axis must be vertical.
5. Make all four welds shown in the drawing below in the vertical welding position.

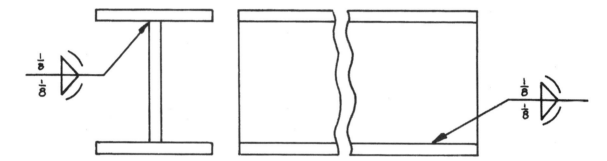

INSPECTION

All welds should be 1/8" (3.2mm) in size, convex, straight, with evenly spaced ripples, and with good fusion at the toes of the weld.

Instructor's initials: _____

Job 6E-3

LEARNING OBJECTIVE

★ You will make a V-groove weld on an outside corner joint in the vertical welding position for this job.

Name_____
Date _____ Class _____
Instructor _____
Grade _____

1. Obtain four pieces 1/8″ × 1 1/2″ × 5″ (3.2mm × 38mm × 127mm) mild steel.

2. Use the following for these welds:

   #_____ drill tip size

   _____ diameter welding rod

   _____ psig oxygen

   _____ psig acetylene

3. Tack weld the weldment shown in step #5. Tack welding may be done in the flat welding position.

4. Place the weldment in a welding fixture or prop it against several firebricks. The weld axis must be vertical.

5. Make all four welds shown in the drawing below in the vertical welding position.

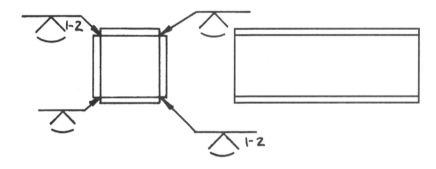

INSPECTION

Each weld bead should be convex, straight, with evenly spaced ripples. The weld should fill the V-groove. Each V-groove weld must have 100% penetration produced by the one-side weld.

Instructor's initials: _____

Lesson 6F

OXYFUEL GAS PROCESSES—WELDING MILD STEEL IN THE OVERHEAD WELDING POSITION

Name _____ Date _____ Lesson Grade _____
Class _____ Instructor _____

LEARNING OBJECTIVE
* You will be able to practice the safety precautions required for overhead welding. You will also be able to weld in the overhead welding position.

INSTRUCTIONS
Carefully read Headings 6.12.3, 6.14, and 6.14.3 of the text. Also study Figures 6-37 and 6-49 in the text. Then complete the following problems.

1. To prevent hot metal from being caught, pant legs should not have _____ on them.

    1. _____

2. When performing the overhead weld, the operator should stand _____, if possible.
    A. above the weld
    B. near the end of the weld
    C. off to one side
    D. None of the above.

    2. _____

3. *True or False?* Overheating will cause the metal to become too fluid and may cause the molten metal to fall.

    3. _____

4. The momentary movement of the flame away from the weld pool and back again is called a _____ motion.

    4. _____

5. The motion in question 4 is used to _____.
    A. superheat the weld pool
    B. cool the weld pool
    C. allow space for the welding rod
    D. allow time to insert the welding rod into the welding pool
    E. create a larger weld pool

    5. _____

6. If the torch tip becomes plugged by molten metal, it should be cleaned using a _____.

    6. _____

7. Whenever the torch is not in your hands, it _____.
   A. should be turned down
   B. should have the acetylene turned off
   C. should be hung on a torch holder
   D. should be placed on the welding table
   E. must be turned off

7. _____

8. What two forces act to keep the molten metal from falling out of the weld pool? _____

9. List four pieces of protective clothing that must be worn to protect the head, hands, body, and feet from burns while overhead welding. _____

10. Define overhead welding. _____

Job 6F-1

LEARNING OBJECTIVE

★ In this job, you will make a square-groove weld on a butt joint, a fillet weld on a lap joint, and a fillet weld on a T-joint.

Name _____
Date _____ Class _____
Instructor _____

1. Obtain five pieces of 1/16″ × 1 1/2″ × 5″ (1.6mm × 38mm × 127mm) mild steel.
2. Use the following for these welds:

    #_____ drill tip size

    _____ diameter welding rod

    _____ psig oxygen pressure

    _____ psig acetylene pressure

3. Tack weld the weldment as shown in step #5. This may be done in the flat welding position.
4. Place the weldment into a positioning fixture. Adjust the positioning fixture so that the joints are high enough to permit overhead welding.
5. The weldment may be repositioned to ensure that each weld is made in the overhead welding position.
6. Make all the welds indicated in the drawing below in the overhead welding position.

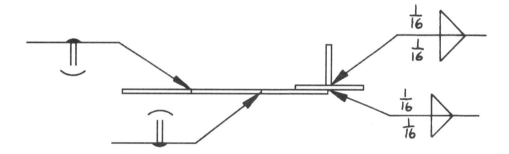

INSPECTION

Check each weld. The butt should have 100% penetration. All welds should be convex, with evenly spaced ripples, and with evidence of good fusion at the toes of the weld.

Instructor's initials: _____
Butt weld grade: _____

Instructor's initials: _____
Lap weld grade: _____

Instructor's initials: _____
T-joint weld grade: _____

Copyright Goodheart-Willcox Co., Inc.

Lesson 7

OXYFUEL GAS CUTTING EQUIPMENT AND SUPPLIES

Name _____ Date _____ Lesson Grade _____
Class _____ Instructor _____

LEARNING OBJECTIVE
* You will be able to identify and correctly use the equipment and supplies used in oxyfuel gas cutting.

INSTRUCTIONS
Carefully read Headings 7.1 through 7.4.2 of the text. Also study Figures 7-1 through 7-23 in the text. Then complete the following problems.

1. Name the parts of the oxyacetylene cutting outfit shown below.

   A. _____
   B. _____
   C. _____
   D. _____
   E. _____
   F. _____
   G. _____
   H. _____
   I. _____
   J. _____
   K. _____
   L. _____

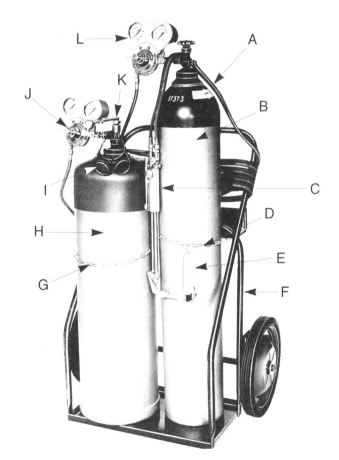

94   Modern Welding Lab Manual

2. What type of cutting torch is shown below in #3? _____
   _____

3. Name the parts indicated in the illustration below.

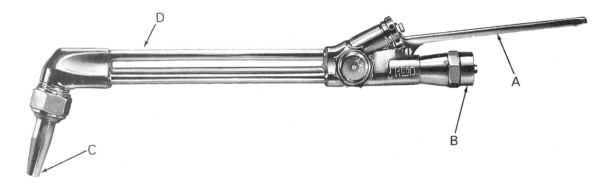

   A. _____
   B. _____
   C. _____
   D. _____

4. Describe the application of the two cutting torch tips shown below.

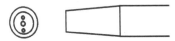

   A. _____

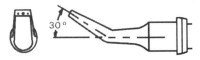

   B. _____

5. What is the name of the tool shown below? _____

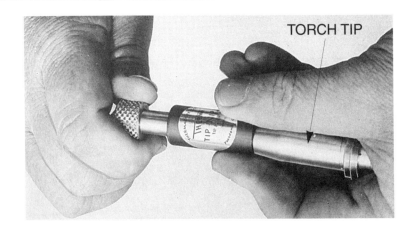

6. What is the name of the tip tool shown below? _____

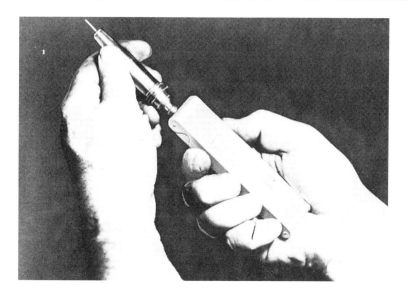

7. The welder below is using a _____ to help cut a circle with the torch.

7.

8. The direction of movement of an electronic pattern tracer is determined by _____.
   A. the thickness of the pattern line
   B. a pinpoint of light traveling through the photoelectric cell
   C. the intensity of the light reflected from the template surface
   D. the speed of the drive motor moving the electronic pattern tracer
   E. the intensity of the light entering the servomotor

8. _____

9. How many servomotors are used to move the rail containing the cutting torches on a cutting machine like the one shown in Figure 7-19 in the text?

9. _____

10. _____ make a heavy duty oxygen regulator the best choice for oxyacetylene cutting.
    A. Two stages and heavy duty springs
    B. High capacity orifice
    C. Permits higher gas flow
    D. Capable of handling a working pressure of 100-150 psig
    E. All of the above.

10. _____

Lesson 8

OXYFUEL GAS CUTTING—CUTTING STEEL

Name _____ Date _____ Lesson Grade _____
Class _____ Instructor _____

LEARNING OBJECTIVE
★ You will be able to cut and gouge various thicknesses of steel.

INSTRUCTIONS
Carefully read Headings 8.1 through 8.10 of the text. Also study Figures 8-1 through 8-30 in the text. Then complete the following problems.

1. What makes it possible for the special oxyacetylene gouging tip to gouge, but not completely cut through, the metal?
   A. The slow forward speed used when gouging.
   B. The tip is held closer to the metal surface when gouging.
   C. A lower velocity (slower-moving) cutting oxygen stream is used.
   D. A smaller cutting oxygen orifice.
   E. The higher flame temperature used when gouging.

   1. _____

2. The nine steps used to start up and light an oxyacetylene cutting station are given below. They are not in the correct order. Place them in the correct order by writing the correct letter in the appropriate blank.

   A. Open the oxygen cylinder valve very slowly until the regulator high pressure gauge reaches its maximum reading. Then turn the cylinder valve all the way open to close the double seating valve. While doing this, the operator should stand to one side of the gauge.

   B. Open the torch oxygen valve one turn. With this valve open, next open the oxygen cutting valve and adjust the oxygen regulator to give the desired operating cutting pressure. If the oxygen cutting valve is not opened while adjusting the oxygen pressure, a drop in oxygen pressure will occur when this valve is opened during the cutting operation. This will result in a reduced preheat flame, and possibly, a poor cut. Close the torch oxygen valve and the oxygen cutting valve. See the table in Figure 8-7 in the text for oxygen and acetylene cutting pressure.

   C. Check the condition of the equipment.

   1. _____
   2. _____
   3. _____
   4. _____
   5. _____
   6. _____
   7. _____
   8. _____
   9. _____

(Continued)

Copyright Goodheart-Willcox Co., Inc.

97

D. Open the acetylene torch valve one turn. Slowly turn in the acetylene regulator adjusting screw until the low pressure acetylene gauge indicates pressure corresponding to tip size. Refer to Figure 8-7 in the text. Close the acetylene torch valve. Regulator pressures are now adjusted. Figure 8-6 in the text shows a single-stage oxygen regulator.

E. Inspect the regulators. Turn the adjusting screws all the way out (regulator closed).

F. Open the acetylene cylinder valve slowly 1/4 to 1/2 turn. Leave the acetylene cylinder valve wrench in place so the cylinder valve may be shut off quickly if necessary.

G. Open the torch acetylene valve until the acetylene flame jumps away from the end of the tip slightly and back again when the torch is given a shake or whipping action. An alternate method of adjusting the acetylene, after the torch is lighted, is to turn on the acetylene until most of the smoke clears from the flame. See Figures 4-5 and 4-6 in the text for illustrations of various conditions of the flame adjustment.

H. To light the torch, open the torch acetylene valve approximately 1/16 turn. Then, use a flint lighter to ignite the acetylene.

I. Next, open the torch oxygen valve and adjust it to obtain a neutral flame. Open the cutting oxygen valve and readjust the preheat flame if necessary. The neutral flame may be altered when the cutting oxygen valve is opened. The torch is now adjusted and is ready to be used as a cutting torch.

3. *True or False?* When cutting cast iron, the metal is preheated and a carburizing flame is used. The casting is cooled slowly to form gray cast iron.

3. _____

4. When using a cutting attachment with the flame extinguished, the _____ valve(s) may be left open if cutting is to continue in a short time.
   A. cutting attachment oxygen preheat
   B. acetylene torch
   C. oxygen and acetylene torch
   D. oxygen preheat valve and the cutting oxygen lever
   E. oxygen torch, the acetylene torch, the oxygen preheat, and the cutting oxygen lever

4. _____

5. The approximate oxygen and acetylene pressure to use when cutting 1″ (25.4mm) mild steel with an equal pressure torch is:

   _____ psig oxygen

   _____ psig acetylene

   _____ kPa

6. What is the suggested angle from vertical and the base metal to tip distance for the cutting tip when cutting heavy metal?
   A. 30°–45° and 1/16″–1/8″ (1.6mm – 3.2mm)
   B. 0°–15° and 1/16″–1/8″ (1.6mm – 3.2mm)
   C. 0°–15° and 1/8″–1/4″ (3.2mm – 6.4mm)
   D. 30°–45° and 1/8″–1/4″ (3.2mm – 6.4mm)
   E. 0°–45° and 1/8″–1/4″ (3.2mm – 6.4mm)

6. _____

7. _____ is generally the cause of a bell-mouth kerf.
   A. Too much acetylene pressure
   B. An oxidizing flame
   C. Too much oxygen pressure
   D. A carburizing flame
   E. Too fast a torch motion

7. _____

8. List seven items of importance that can be changed or varied to possibly improve the quality of the cut. ___

9. List three changes which can be made in order to correct an excessive lag or drag when oxyfuel gas cutting.

10. List four cutting methods which have been automated by the use of automatic cutting machines. _____

11. *True or False?* When oxyfuel gas cutting, caution should be taken because considerable sparking and flying of sparks (globules of molten metal), accompany cutting operation.

11. _____

12. *True or False?* Tanks and containers should only be welded or cut by an experienced welder.

12. _____

Job 8-1

LEARNING OBJECTIVE ★ In this job, you will learn to cut metal using the oxyfuel gas cutting torch.	Name _____ Date _____ Class _____ Instructor _____ Grade _____

1. Obtain a piece of mild steel 1/4″ × 8″ × 8″ (6.4mm × 206mm × 206mm).

2. Obtain a second piece which is thicker than 1/4″ (6.4mm), preferably 1″ (25.4mm) thick. The thick piece should be at least 3″ (76mm) wide.

3. Using a soapstone, mark eight 1″ (25.4mm) wide strips on the 1/4″ (6.4mm) thick plate.

4. Place the 1/4″ (6.4mm) plate on the cutting table. The plate must be placed on the table so that the cutting table is not damaged as each cut is made. The plate may be placed on firebricks. The area to be cut should hang over the edge of the firebrick.

5. While performing these cuts with a positive-pressure torch, the following will be used:

   _____ preheating orifice size

   _____ cutting oxygen orifice size

   _____ psig (_____ kPa) oxygen pressure

   _____ psig (_____ kPa) acetylene pressure

   _____ in./min. (_____ mm/min.) cutting speed

6. Cut four pieces 1″ (25.4mm) wide with square edges. Cut the remaining four pieces so that three have a 45° bevel or chamfer on the edges.

7. Place the thicker plate on the cutting table. Arrange the plate so the table is not damaged.

8. The plate thickness used is _____″ (_____ mm) thick.   8. _____

9. While performing the cut on this thicker metal, the following will be used:

   _____ preheating orifice drill size

   _____ cutting oxygen orifice drill size

   _____ psig oxygen pressure _____ kPa

   _____ psig acetylene pressure _____ kPa

   _____ in./min. cutting speed _____ mm/min.

10. Using a soapstone, mark off three pieces 1″ (25.4mm) wide.

11. Cut one piece with a square edge and one piece with a 45° bevel or chamfered edge.

Instructor's initials: _____

Job 8-2

LEARNING OBJECTIVE

✱ In this job, you will learn to cut shapes using the manual oxyfuel gas torch.

Name _____
Date _____ Class _____
Instructor _____
Grade _____

1. Obtain two pieces of plain carbon steel that measures 5/16" × 2" × 6" (7.9mm × 50.8mm × 152.4mm).

2. Using a soapstone, draw the shape shown by the dimensions on the drawing below:

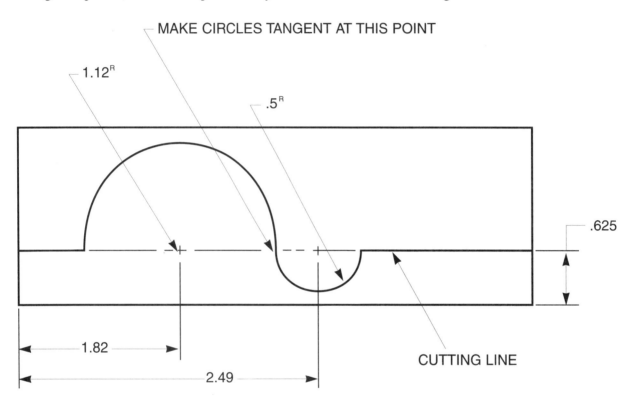

3. Place one piece of metal on the cutting table and prepare the cutting outfit to cut the shape drawn on the base metal.

4. While making the cuts with a positive-pressure cutting torch, the following variables will be used:

   _____ preheating orifice size

   _____ cutting oxygen orifice size

   _____ psig (_____ kPa) oxygen pressure

   _____ psig (_____ kPa) acetylene pressure

   _____ in./min. (_____ mm/min.) cutting speed

5. Cut the shape drawn on the base metal.

6. Cut the second piece of metal in the same manner.

Instructor's initials: _____

Job 8-3

LEARNING OBJECTIVE	
✷ In this job, you will learn to remove the weld reinforcement from the face of a weld.	

Name _____
Date _____ Class _____
Instructor _____
Grade _____

1. Obtain three previously made welds or beads from the scrap metal container and have the instructor approve your choices.

2. Obtain a gouging tip with two holes and a #70 preheat orifice size and install it on the cutting torch of your cutting outfit.

3. While performing this gouging operation, you will use and set the following variable on your cutting station:

    _____ preheating orifice size

    _____ cutting oxygen orifice size

    _____ psig (_____ kPa) oxygen pressure

    _____ psig (_____ kPa) acetylene pressure

4. Remove the weld reinforcement from one of the three welds.

5. Check the appearance of the base metal surface to determine if any changes are required to the oxygen and acetylene pressure, your technique, or travel speed.

6. Remove the weld reinforcement from the other two welds.

Instructor's initials: _____

Job 8-4

LEARNING OBJECTIVE

★ In this job, you will learn to set up and inspect an oxyfuel gas cutting straight line track burner.

Name _____
Date _____ Class _____
Instructor _____
Grade _____

1. Check that the track used with your OFC electrically driven cutting torch carriage (track burner) is laying flat on the cutting table.

2. Place the cutting torch carriage on the track. Be certain that all the wheels are in the tracks.

3. Place the base metal to be cut next to the carriage on the cutting table. Be certain that it is laying flat.

4. Mark the line to be cut.

5. Adjust the angle of the cutting torch by loosening the clamp that holds the torch on the carriage. A square cut or a chamfer of any angle may be made. Tighten the torch after the torch angle to base metal is correct.

6. Adjust the cutting torch in or out to align the torch tip with the cutting line.

7. Adjust the height of the cutting tip above the base metal.

8. After these adjustments are made, run the carriage along the track to ensure that the torch and hoses are not hitting anything as the cut progresses along the cutting line.

9. Using the proper size open end wrench, check that all hose fittings are tight on the carriage, torch, and regulators. **Caution:** Do not overtighten the brass fittings.

10. Check the hoses carefully for worn or cut areas.

11. Turn on the cutting outfit and set the oxygen and acetylene working pressures on the regulators. See Heading 8.4.1 in the text.

12. Set the forward travel speed on the torch carriage and engage the clutch to test the travel speed.

13. The cutting torch carriage and torch are now ready to make the cut.

14. Have the instructor check your inspection and machine set-up.

Instructor's initials: _____

Job 8-5

LEARNING OBJECTIVE

★ In this job, you will learn to make straight line cuts on plain carbon steel to produce square and bevel edges.

Name _____
Date _____ Class _____
Instructor _____
Grade _____

1. Obtain two pieces of plain carbon steel that measure 1/2" × 8" × 20" (12.7mm × 203mm × 508mm).

2. Using a soapstone, mark four 2" (50.4mm) wide strips on each piece.

3. To cut this thickness of metal, using a positive-pressure type cutting torch, the following variables must be used:

   _____ preheating orifice size

   _____ cutting oxygen orifice size

   _____ psig (_____ kPa) oxygen pressure

   _____ psig (_____ kPa) acetylene pressure

   _____ in./min. (_____ mm/min.) cutting speed

4. Place the base metal on a cutting table so that it lays flat.

5. Set up the electric cutting torch carriage using the settings found above in #3. Test the travel of the carriage.

6. Set the torch angle to cut a square edge in the kerf.

7. Light the cutting torch in the same manner used to light an OFW torch.

8. Engage the carriage clutch to begin the cut.

9. Cut the first piece on the three marked places.

10. Watch the cut as it progresses and stop the carriage, by disengaging the carriage clutch, if the variables need to be changed to make a better cut.

11. Place the second piece of base metal on the cutting table and align the carriage and cutting torch.

12. Align the carriage and torch with the first line to be cut.

13. Set the torch so that the cutting tip will cut a kerf at a 45° angle.

14. Examine the first bevel cut and change any of the cutting variables to create an acceptable cut surface on the next two cuts.

15. Cut the next two bevels on the cutting lines marked.

Instructor's initials: _____

Lesson 9

SOLDERING

Name _____ Date _____ Lesson Grade _____
Class _____ Instructor _____

> **LEARNING OBJECTIVE**
>
> ✶ You will be able to describe the principles of soldering. You will also be able to demonstrate various methods of soldering.
>
> **INSTRUCTIONS**
>
> Carefully read Headings 9.1 through 9.15 of the text. Also study Figures 9-1 through 9-25 in the text. Then complete the following problems.

1. Name three items that must be worn while working with cleaning solvents, acids, and pickling solutions. __
   _____
   _____
   _____

2. The melting temperature of an alloy of both metal A (500°F [260°C]) and metal B (400°F [204°C]), will melt _____.
   A. below the melting temperature of metal A
   B. below the melting temperature of metal B
   C. at a temperature equal to the sum of metal A's and metal B's melting temperature
   D. at a temperature above metal A's melting temperature
   E. at a temperature above metal B's melting temperature

   2. _____

3. What is the eutectic alloy percentage and the liquidus temperature for a tin-lead solder?

   A. _____ % lead

   B. _____ % tin

   C. _____ °F liquidus temperature

   D. _____ °C liquidus temperature

4. What type solder is recommended for soldering stainless steel food handling tables and equipment?

   4. _____

5. What solder alloying element is a health hazard if inhaled and must be used only when excellent ventilation is possible?

   5. _____

Copyright Goodheart-Willcox Co., Inc.

6. Identify the purpose(s) for a good soldering flux. 6. _____
   A. Clean the metal surfaces.
   B. Prevent metal oxidation.
   C. Provide a good wetting action.
   D. Clean the welding torch.
   E. Remove oil, grease, and paint.

7. A(n) _____ flux cleans the best. 7. _____
   A. organic
   B. nonactive rosin
   C. rosin
   D. inorganic
   E. None of the above.

8. What type of flux is corrosive during the soldering operation, noncorrosive after soldering, and ideal for use when soldering electronic assemblies? 8. _____

9. _____ rosin flux cleans best. 9. _____
   A. Nonactive
   B. Fully active
   C. Mildly active

10. Which of the following steps should be done to remove a very corrosive inorganic flux? 10. _____
    A. Wash with hot water containing 2% hydrochloric acid.
    B. Wash with hot water and a mild detergent.
    C. If needed, wash with hot water and sodium carbonate.
    D. Rinse with hot water.
    E. All of the above.

11. Name the following types of soldered joints.
    A. _____
    B. _____
    C. _____
    D. _____
    E. _____
    F. _____

12. The process of adhering a very thin layer or film of solder to a metal surface prior to soldering is called _____. 12. _____

13. Torch soldering is generally done using the same seven procedural steps. Below are the suggested steps for torch soldering. The steps are out of the proper order. Place the letters in the correct order in the spaces to the right.

    A. Support the joint through the soldering and cooling period.

    B. If the solder does not adhere, stop the operation and reclean and reflux the joint.

    C. Clean the surfaces to be soldered. Refer to Heading 9.1.1 in the text.

    D. Continue to move the torch close to the joint so the joint does not overheat. Touch the solder to the joint. Continue to add solder to the joint until it is completely filled.

    E. A common soldering error is to add solder to a cool joint by melting the solder with the torch. The base metal should melt the solder.

    F. Heat the joint with the torch until the joint is just hot enough to melt the solder alloy used.

    G. Apply the correct flux to the joint surfaces. Refer to Heading 9.3 in the text.

    1. _____
    2. _____
    3. _____
    4. _____
    5. _____
    6. _____
    7. _____

14. The principle of _____ action is used to draw solder into the small clearance space between the pipe and fitting in a pipe joint.

    14. _____

15. What method of soldering is shown in the figure below? _____

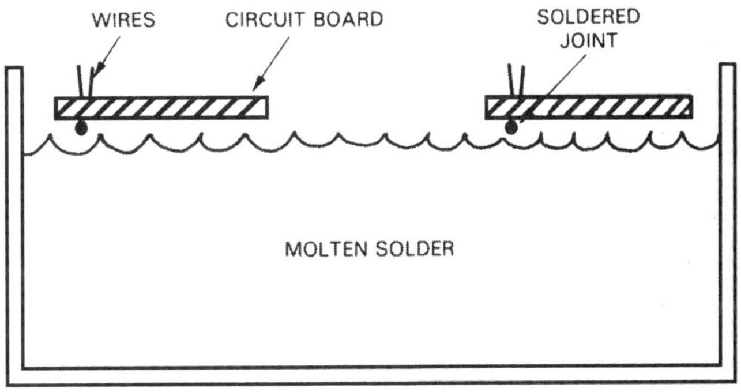

16. Which of the methods of soldering listed below generally use pre-formed and pre-placed solder forms in the assembly to be soldered?
    A. Wave soldering
    B. Oven or infrared soldering
    C. Torch soldering
    D. Resistance and induction soldering
    E. Dip soldering

    16. _____

17. _____ flux is recommended for use when soldering stainless steel.
    A. Organic
    B. Rosin
    C. Inorganic
    D. Mildly active rosin
    E. Nonactive rosin

    17. _____

18. Toxic (unhealthy and possibly deadly) fumes result from heating which of the following metals?
    A. Cadmium
    B. Beryllium
    C. Stainless steel
    D. Mild steel
    E. Copper

18. _____

19. List eight procedures that must be performed to successfully produce a soldered joint. _____

_____

_____

_____

_____

_____

_____

_____

_____

_____

_____

20. _____ solder is recommended for soldering gold.
    A. Corrosive
    B. Noncorrosive

20. _____

Job 9-1

LEARNING OBJECTIVE

★ In this job, you will learn to solder copper fittings for plumbing drain lines using a tin-lead solder. **Note:** Tin-lead solders must not be used on plumbing that will carry drinking water.

Name _____
Date _____ Class _____
Instructor _____
Grade _____

1. Obtain the following copper fittings:
   1/2" cap
   1/2" elbow
   1/2" union (female half only)

Note: If the fittings have already been assembled, unsolder them.

2. Obtain two lengths of 1/2" copper tubing approximately 7" long.

3. Use a corrosive or noncorrosive flux and a common 50–50 tin-lead solder.

4. Using an air-acetylene or propane torch and the correct soldering procedure, assemble the pipe and fittings as shown below:

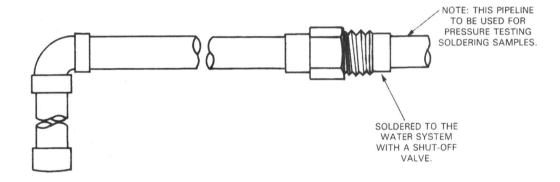

INSPECTION

Assemble your soldered assembly to the waterline with the union. Open the water valve and test for leaks. Turn off the water valve and remove your assembly.

Instructor's initials: _____

Job 9-2

LEARNING OBJECTIVE

★ You will learn to solder a folded metal seam in this job.

Name _____
Date _____ Class _____
Instructor _____
Grade _____

1. Obtain two pieces of 28 ga. galvanized steel measuring 1 1/2″ × 6″ (38mm × 152mm).

2. Form seams on each piece to make the following assembly.

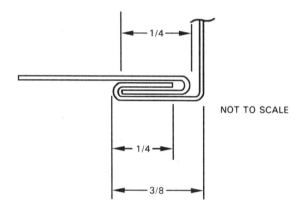

3. Clean the piece. Then assemble and solder the joint. Use the correct soldering procedure. The solder should be a common 50–50 tin-lead solder. Use a corrosive or noncorrosive soldering flux.

INSPECTION

There should not be evidence of the use of too much solder. A thin film of solder should be visible in all folded seams.

Instructor's initials: _____

Lesson 10A

BRAZING AND BRAZE WELDING

Name _____ Date _____ Lesson Grade _____
Class _____ Instructor _____

LEARNING OBJECTIVE
✶ You will be able to define brazing and braze welding. You will also be able to choose the correct flux and brazing filler metal.

INSTRUCTIONS
Carefully read the introduction to Chapter 10 and Headings 10.1 through 10.4 of the text. Also study Figures 10-1 through 10-8 in the text. Then complete the following problems.

1. What is the difference between brazing and braze welding?   1. _____
   A. Braze welding is done above 840°F (450°C).
   B. Brazing is done with heavy layers of brazing filler metal applied.
   C. Braze welding does not rely on capillary action to distribute the brazing filler metal.
   D. Brazing is generally done on thick metal sections.
   E. Brazing is done above 840°F (450°C).

2. List two reasons why parts are brazed or braze welded rather than welded. _____
   _____
   _____

3. The most important thing to do before applying the flux prior to   3. _____
   braze welding or brazing is to _____ the metal surfaces.

4. Name six ingredients that may be in a brazing flux. _____
   _____
   _____
   _____
   _____
   _____

5. Name the six braze welded joints shown below.

   A. _____
   B. _____
   C. _____
   D. _____
   E. _____
   F. _____

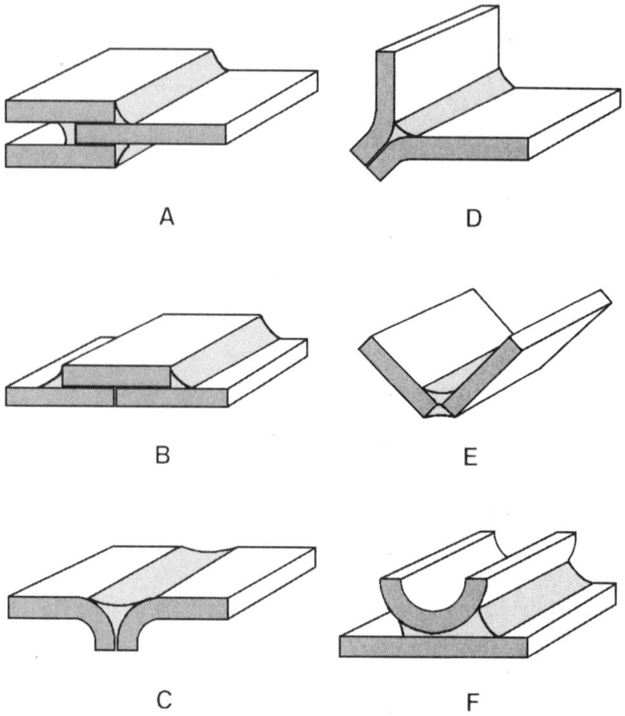

6. through 8. When brazing a nickel based alloy containing small amounts of aluminum (Al) or titanium (Ti), the following questions must be answered.

   6. What AWS brazing flux type number is used? 6. _____

   7. What filler metal is recommended for all applications? 7. _____

      What filler metal is recommended for copper based alloys only? _____

   8. What are the flux ingredients? _____
   _____
   _____

9. List eight criteria that must be considered when choosing a brazing flux: _____
   _____
   _____
   _____
   _____
   _____
   _____
   _____
   _____

10. *True or False?* Fluxes used for braze welding must withstand higher temperatures for longer periods than are required of brazing fluxes.   10. _____

Job 10A-1

LEARNING OBJECTIVE	Name_____
★ In this job, you will braze weld the following types of joints in the flat welding position: butt, lap, and T-joint.	Date _____ Class _____ Instructor _____ Grade _____

1. Obtain six pieces of mild steel that measure 1/8″ × 1 1/2″ × 5″ (3.2mm × 38mm × 127mm).

2. Clean the joint areas of all pieces at least 1/2″ (12.7mm) back from the joint.

3. Using a BCu or copper brazing filler metal, what flux ingredients are suggested in Figure 10-8 in the text?

   _____
   _____

4. Braze weld two of each of the following joints.

Note: Tack braze each joint at three points to hold in position.

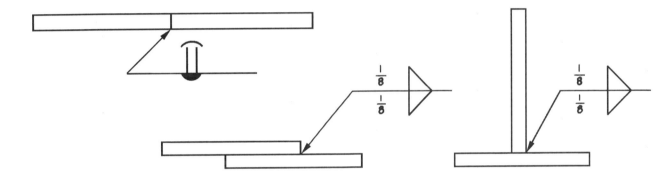

INSPECTION

The braze-welded beads should be straight, with an even bead width.

Instructor's initials: _____

Lesson 10B

BRAZING AND BRAZE WELDING

Name _____ Date _____ Lesson Grade _____
Class _____ Instructor _____

LEARNING OBJECTIVE
★ You will be able to braze stainless steel, cast iron, and some nonferrous metals. You will also be able to determine which of the various fluxes to use for each metal.

INSTRUCTIONS
Carefully read Headings 10.4.1 through 10.8 of the text. Also study Figures 10-9 through 10-40 in the text. Then complete the following problems.

1. When selecting a brazing filler metal, the _____ must be considered.   1. _____
   A. brazing and service temperature required for the brazement
   B. compatibility with the base metals
   C. method of heating used
   D. All of the above.

2. List the four brazing alloys suggested for brazing copper to copper. (See Figure 10-13 in the text.)
   _____
   _____
   _____
   _____

3. *True or False?* Brazing nickel (Ni) to molybdenum (Mo) is not recommended.   3. _____

4. List the four brazing filler metals suggested to join tool steel to carbon steel. _____
   _____
   _____
   _____

5. When alloys of silver are used to join metals, it is called silver _____ because the process is done at temperatures above 840°F (450°C).   5. _____

Copyright Goodheart-Willcox Co., Inc.

117

6. *True or False?* Properly made silver brazed joints are strongest when they are thinner than 0.002″ (.05mm).

6. _____

7. List at least six heat sources used for brazing. _____

_____

_____

_____

_____

_____

8. The brazing filler metal will not flow over the base metal surface unless the _____ metal surface is heated to the brazing filler metal flow temperature.

8. _____

9. When silver brazing, the flux has a clear (watery) appearance at _____ °F (_____°C). At this point, the filler metal should be added.
   A. 212°F (100°C)
   B. 600°F (316°C)
   C. 1100°F (593°C)
   D. 1370°F (743°C)
   E. 1230°F (662°C)

9. _____

10. Indentify the steps that may be taken to clean magnesium (Mg) after brazing is done.
    A. Clean in hot running water.
    B. Scrub with strong chemicals.
    C. Mechanically scrub in water.
    D. Polish until bright.
    E. Dip into a chrome pickle for about two minutes.

10. _____

Job 10B-1

LEARNING OBJECTIVE	Name _____
★ In this job, you will learn to braze weld a V-groove on an outside corner joint in the flat welding position.	Date _____ Class _____ Instructor _____ Grade _____

1. Obtain three pieces of carbon steel that measure 1/4" × 1 1/2" × 5" (6.4mm × 38mm × 127mm).

2. Braze weld the pieces into the shape shown in the drawing below:

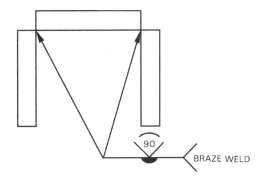

3. If no silver, gold, nickel, or zinc based filler metal is available, what filler metal would you use? (See Figure 10-13 in the text.)

4. What ingredients may be in the flux that you use? (See Figure 10-8 in the text.) _____

_____
_____
_____
_____
_____
_____

INSPECTION

The braze welding bead should be convex. The bead should not go beyond the edges of the groove more than about 1/16" (1.6mm). The ripples in the bead should be evenly spaced and the bead should have a constant width. Complete penetration should be indicated over the entire length of the joint.

Instructor's initials: _____

Lesson 11A

SHIELDED METAL ARC WELDING EQUIPMENT AND SUPPLIES

Name _____ Date _____ Lesson Grade _____
Class _____ Instructor _____

LEARNING OBJECTIVE
★ You will be able to describe the various types of shielded metal arc welding (SMAW) machines. You will be able to determine the various accessories and supplies required in SMAW. You will also be able to select the correct current and electrode.

INSTRUCTIONS
Carefully read Headings 11.1 through 11.11.3.1 of the text. Also study Figures 11-1 through 11-31 in the text. Then complete the following problems.

1. A complete SMAW station consists of _____.
   A. an ac or dc arc welding machine
   B. an electrode holder
   C. an electrode lead
   D. a workpiece lead
   E. a workpiece clamp
   F. All of the above.

   1. _____

2. An arc welding machine which consists of a dc rectifier and an alternator is a(n) _____ machine.
   A. positive-voltage
   B. direct current
   C. ac/dc
   D. alternating current
   E. None of the above.

   2. _____

3. *True or False?* The general slope of the volt-ampere curve is called the output slope of the power source.

   3. _____

Copyright Goodheart-Willcox Co., Inc.

121

4. Which of the following describes the volt-amperage curve shown below?
   A. Constant current
   B. Shallow slope
   C. Steep slope
   D. Drooper
   E. Constant voltage

4. _____

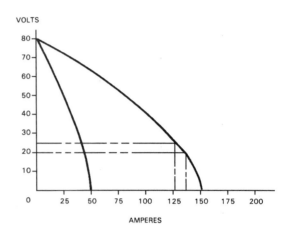

5. The "drooper" arc welding machines are the _____ type.
   A. ac
   B. dc
   C. constant current
   D. constant voltage
   E. ac/dc

5. _____

6. The constant _____ machine is most desirable when doing manual arc welding.

6. _____

7. *True or False?* A manual arc welder would normally prefer a flatter, constant current output slope when welding in a horizontal, vertical, or overhead position.

7. _____

8. List the three principal components of the transformer type arc welder. _____

_____

_____

_____

9. What type of control is shown in the schematic below? _____

_____

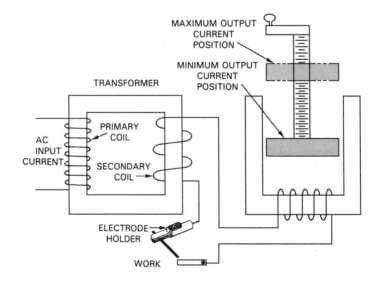

10. The symbol for the device which changes ac to dc by only allowing current to flow in one direction is shown below. What is it called?

10. _____

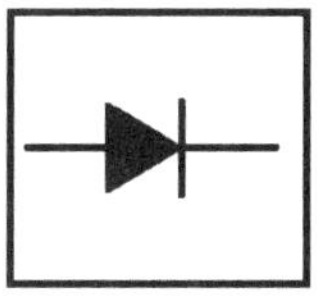

11. What do the letters DCEN mean? _____
_____

12. DCSP has the same direction of current flow as DCE _____. 12. _____

13. If a welder varies the arc gap (voltage) of the system and only a very small change in the delivered current occurs, the welding machine is called a constant _____ arc welding machine. 13. _____

14. What do the letters DCEP mean? _____
_____

15. What is the Rated Load (welding) Voltage for a Class I, 400 ampere, arc welding machine? _____
_____

Show your calculations below:

16. Describe the term "duty cycle." _____
_____
_____
_____

17. *True or False?* Arc welding cables should be kept as short as possible. They should also be kept as far away from each other as possible. 17. _____

18. A duty cycle–welding amperage chart is shown in Figure 11-31 of the text. How many minutes out of every 10 minutes can this machine operate continuously at 125 amperes? 18. _____

19. What NEMA arc welding machine classification is assigned to a machine that can deliver its rated output at a constant duty cycle of 30%- 50%? 19. _____

20. What duty cycle is recommended for automatic and semiautomatic arc welding operations? 20. _____
    A. 20
    B. 40
    C. 60
    D. 80
    E. 100

Lesson 11B

SHIELDED METAL ARC WELDING EQUIPMENT AND SUPPLIES

Name _____ Date _____ Lesson Grade _____
Class _____ Instructor _____

LEARNING OBJECTIVE
★ You will be able to select the proper size electrode and work leads. You will be able to select the correct electrode type and electrode diameter for a given job. You will also be able to select welding lens shades.

INSTRUCTIONS
Carefully read Headings 11.4 through 11.12 of the text. Also study Figures 11-31 through 11-62 in the text. Then complete the following problems.

1. Carefully study Figure 11-34 of the text. What size welding leads are required to deliver 200 amperes to a job 80′ from the arc welding machine?

    1. _____

2. Insulated and uninsulated terminals are connected to the welding leads by what four methods? _____
   _____
   _____
   _____
   _____

3. Name the item shown below. _____

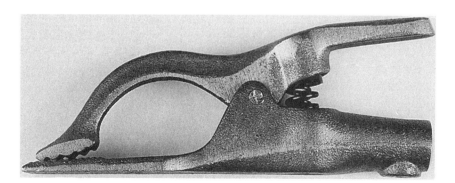

Copyright Goodheart-Willcox Co., Inc.

125

4. Which of the following are actions and purposes of the electrode flux coating?
   A. Forms CO or H_2 or prevents oxygen from reaching the molten metal.
   B. Helps produce a weld with excellent chemical and physical properties.
   C. Forms a thick coating which it cools to protect the molten metal.
   D. Creates a cleaning of fluxing action to remove unwanted materials from the molten weld metal.
   E. All of the above.

4. _____

5. *True or False?* The materials in a covered electrode determine the polarity and electrical characteristics of the electrode.

5. _____

6. Electrode specification _____ describes electrodes for use on aluminum and aluminum alloys.
   A. AWS A5.1
   B. AWS A5.2
   C. AWS A5.3
   D. AWS A5.11
   E. AWS A5.15

6. _____

Answer questions 7 through 11 using the electrode number **E9013-A1**.

7. What is the tensile strength of this electrode? _____

8. Name the covering composition of this electrode. _____

9. In what position(s) can this electrode be used? _____

_____

_____

10. What does "**A1**" mean? _____

11. With what current may this electrode be used? _____

12. What type current is required to weld with an E6010 electrode? _____

13. How does the welder know what electrode he or she is using? _____

_____

_____

_____

14. _____ more filler metal is deposited with an E7028 iron powder electrode than is deposited with the same diameter E6012 electrode.
    A. 0%
    B. 22%
    C. 45%
    D. 78%
    E. 100%

14. _____

15. What polarity is used with a chromium or chromium-nickel steel electrode with the number E310-16?

15. _____

16. How many electrodes are contained in the electrode specification AWS A5.3-88?

16. _____

17. To eliminate the time required to travel back and forth to the welding machine, several manufacturers provide _____ _____ devices which may be kept near the operator for convenient control of the machine.

17. _____

18. Name the item shown below: _____

19. Identify the item shown below and describe its function.

_____
_____
_____
_____
_____

20. Welding filter _____ is suggested for use in shielded metal arc welding on 1/4″ (6.4mm) metal.
    A. 10
    B. 11
    C. 12
    D. 13
    E. 14

20. _____

Lesson 11C

ARC WELDING SAFETY

Name _____ Date _____ Lesson Grade _____
Class _____ Instructor _____

LEARNING OBJECTIVE
★ You will be able to practice the safety procedures and precautions required when working with various arc welding processes such as SMAW, GTAW, and GMAW.

INSTRUCTIONS
Carefully read Headings 11.10, 11.11, 12.4.1, 12.4.2, 12.5, 12.7.5, 12.7.8, 12.8, and 13.4.1 of the text. Review Headings 5.2.1, 6.5.2, 6.5.5, 6.14.1, 6.14.2, 13.2, 13.20, 13.22, 14.4, 14.15, 15.4, 15.7, and 15.15 of the text. Then complete the following problems.

1. The line voltage to an industrial arc welding machine is generally 220 V or _____ V. Because of this, only an experienced electrician should work on the electrical power connections used in an arc welding machine.

    1. _____

2. When doing SMAW on 1/4″ (6.4mm) metal for long periods of time, a # _____ filter lens should be used. Also, a pair of _____ goggles should be worn to reduce eye damage from flashes behind the helmet.

    2. _____
       _____

3. The inlet to the ventilation pickup duct should be located so that fumes are removed before they reach the _____ _____.

    3. _____

4. Which of the following must be clean and tight to reduce electrical resistance?
    A. Booth curtains.
    B. Electrode holder jaws.
    C. Workpiece lead at the work.
    D. Workpiece lead at the electrode holder.
    E. Electrode lead at the machine and holder.

    4. _____

Copyright Goodheart-Willcox Co., Inc.

5. Name five hazards to avoid when arc welding. _____
_____
_____
_____
_____
_____

6. *True or False?* You should avoid using arc welding equipment in damp places or with damp welding gloves.

6. _____

7. Select the one statement below that is not important when working around arc welding equipment.
   A. Never look at the arc from any distance without wearing an approved filter lens.
   B. All pockets should be covered.
   C. Do not carry plastic pens, combs, or matches in your pockets.
   D. Never work without adequate ventilation.
   E. Take every precaution to eliminate H_2O and O_2.

7. _____

8. An arc welding machine must never be started under load. Because of this, the electrode holder must always be hung on a(n) _____ hanger before the machine is started.

8. _____

9. *True or False?* The welder should always wear safety goggles when cleaning metals.

9. _____

10. List six items that a welder should wear when arc welding in the overhead position. _____
_____
_____
_____
_____
_____
_____

11. Why should you never use homemade or non-NEMA approved transformer equipment? _____
_____
_____
_____

12. With the exception of hydrogen (H_2), the shielding gases used in GMAW and GTAW are not combustible. What makes handling these cylinders of gas dangerous? _____
_____
_____

13. Which of the following statements is not important when the safe handling of welding shielding gas cylinders is involved?
    A. When storing or moving cylinders, the safety cap must be screwed on securely.
    B. When in use or stored, cylinders must be securely fastened to a stable object.
    C. Care must be taken to avoid accidentally damaging cylinders with an arc or cutting torch.
    D. Cylinders should always be used and stored in the upright (vertical) position.
    E. Cylinders must be changed when they drop below 250 psig or 1724 kPa.

13. _____

14. *True or False?* Oxygen and shielding gas cylinders should not be used in the partly open position.

14. _____

15. *True or False?* If a label on a cylinder is not readable or missing, do not assume that a cylinder contains a particular gas. The cylinder should be returned to the supplier.

15. _____

16. Shielding gases are used with a pressure regulator. The regulator screw should be _____ before the cylinder valve(s) are opened.
    A. screwed all the way in until the threads feel tight
    B. screwed out until the threads feel loose
    C. just 1/4 turn open
    D. half-open
    E. The position does not matter.

16. _____

17. List at least four metals that, when welded, create fumes that are irritating or toxic. _____

_____

_____

_____

_____

18. Do not stand in front of the _____ as the cylinder valve is opened, even if it is opened slowly.

18. _____

19. Which two toxic gases do ventilation and smoke extractors remove from the welding area when GMAW?

_____

_____

20. When doing GMAW on ferrous metals using up to a 5/32" (4mm) diameter electrode, what size filter lens should be used?
    A. 6
    B. 8
    C. 10
    D. 11
    E. 12

20. _____

21. If a film of chlorinated hydrocarbon solvent is left on metal after cleaning, _____ gas will be formed when the metal is arc welded.

21. _____

22. Unless you are wearing supplied air breathing equipment, death from suffocation can occur if you enter an area which is filled with a _____ gas.

22. _____

23. Compressed air may be used to clean out the empty wire feed tube when GMAW. When this is done, be certain that the tube is pointed _____.

24. All arc welding should be done in a booth or area shielded by curtains to protect others from arc _____ or _____.

24. _____
   _____

25. Do not touch any _____ electrode holder or GTAW torch with bare skin or wet gloves. A fatal shock may result.
    A. new
    B. large diameter
    C. uninsulated
    D. insulated
    E. old-fashioned

25. _____

Lesson 12A

ARC WELDING FUNDAMENTALS

Name _____ Date _____ Lesson Grade _____
Class _____ Instructor _____

LEARNING OBJECTIVE
★ You will be able to describe the fundamentals of the shielded metal arc welding (SMAW) process. You will also be able to strike an arc, run a bead, and "read a bead."

INSTRUCTIONS
Carefully read Headings 12.1 through 12.5.5 of the text. Also study Figures 12-1 through 12-27 in the text. Then, complete the following problems.

1. The temperature of the arc in SMAW is _____ °F (_____ °C) to _____ °F (_____ °C).   1. _____

2. What is the polarity of the welding circuit shown below?   2. _____

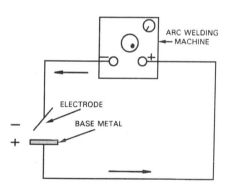

3. In a constant current machine, if the arc gap is increased, the resistance will _____ and the voltage output of the machine must _____.   3. _____

4. The decision to use DCEN or DCEP often depends on such variables as the metal thickness, the position of the joint, the depth of penetration desired, the type of base metal and _____.

5. Electrodes intended for use with ac have _____ agents in the covering to help _____ the arc.   5. _____

Copyright Goodheart-Willcox Co., Inc.

6. Name the various parts of the arc weld and electrode shown below.

   A. _____
   B. _____
   C. _____
   D. _____
   E. _____
   F. _____
   G. _____
   H. _____

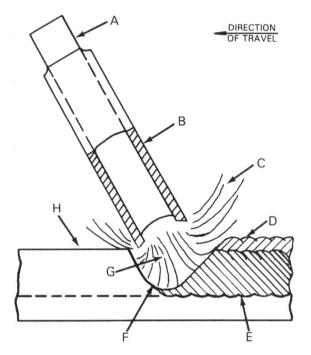

7. Before making a through inspection of the arc welding station, the arc welding machine should be turned _____ to check it properly.

   7. _____

8. Assuming that a 5/32" (4mm) diameter electrode is used, answer the following questions.

   A. What is the maximum width of a stringer bead?

   B. What is the maximum width of a weaving bead?

   C. What is the electrode angle forward of vertical?

   8. A. _____
      B. _____
      C. _____

9. Name the two methods for striking (starting) the shielded metal arc shown below.

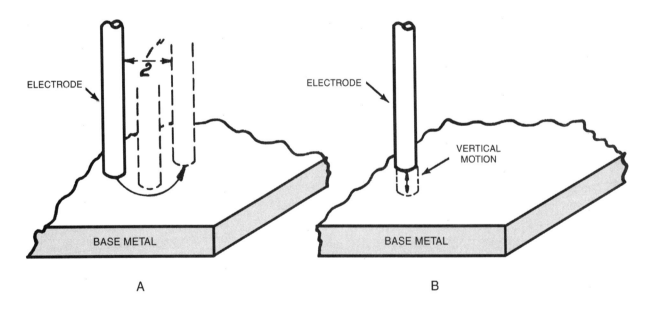

   A. _____
   B. _____

Lesson 12A   Arc Welding Fundamentals   **135**

10. List five factors which the welder must control or select in order to produce a good weld. _____

    _____
    _____
    _____
    _____
    _____

11. List the four factors which an arc welder must control in order to produce an acceptable bead. _____

    _____
    _____
    _____
    _____

12. Much can be told by watching the arc bead as it is being made and when it is completed. From the appearance of the beads below, describe the conditions which caused the bead shapes shown.

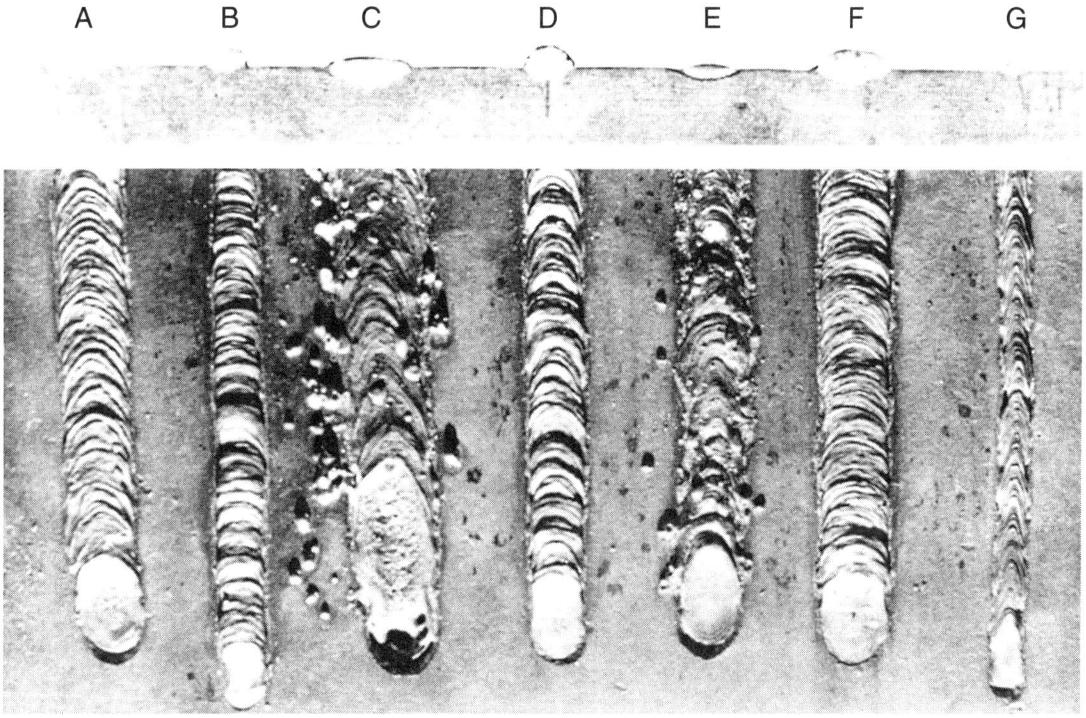

    A   B   C   D   E   F   G

    A. _____
    B. _____
    C. _____
    D. _____
    E. _____
    F. _____
    G. _____

13. Study the illustration below. Explain what must be done in each of the three steps.

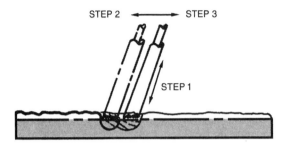

Step 1: _____

Step 2: _____

Step 3: _____

14. The slag coating must be removed before a second bead is placed on top of the first bead. This procedure should prevent slag _____.

14. _____

Job 12A-1

LEARNING OBJECTIVE
✷ In this job, you will demonstrate your ability to thoroughly inspect an SMA, GMA, FCA or GTA welding station, or a PAC station as assigned by the instructor.

Name _____
Date _____ Class _____
Instructor _____
Grade _____

1. Be certain the machine is turned off. Proceed with your inspection. Place an (X) in the "OK" column if the item checked is in good condition. Place an (X) in the "Needs Repair" column if the item needs repair.

	OK	Needs Repair

2. Check the work lead and electrode lead for:

   A. Tight connections at the machine.

   B. Tight connections at the ground clamp or workpiece lug.

   C. Tight connections at the electrode holder or gun.

   D. Cuts, worn areas, or deep cracks over the full length of each lead.

   E. Tight water and gas hose connections.

   F. Cuts, worn areas, or deep cracks over the full length of each hose.

3. Be certain that a fully insulated electrode hanger is located near the work table.

4. Turn on the ventilation system. While holding a large piece of paper, check to determine that the ventilation system is pulling air from the booth.

5. Check the walls of the booth and curtains for holes that may need repair.

6. Make certain that the pliers, chipping hammer, and wire brush which you may use are in good working condition.

7. Check your arc helmet for holes or cracks.

8. Check that your filter lens is the correct number and that the cover lens is clean and in good condition.

Instructor's initials: _____

Job 12A-2

> **LEARNING OBJECTIVE**
> ★ In this job, you will practice striking the arc and running short beads.

Name _____
Date _____ Class _____
Instructor _____
Grade _____

1. Obtain a piece of mild steel which measures 1/4″ × 6″ × 6″ (6.4mm × 152mm × 152mm) and four 1/8″ diameter, E6012 electrodes to begin this job. **Note:** This plate will be used for several jobs, so do not lose it.

2. Using a rule and a piece of soapstone or chalk, draw lines 3/8″ (9.5mm) apart across the full 6″ (152mm) length of the metal on both sides. See figure below.

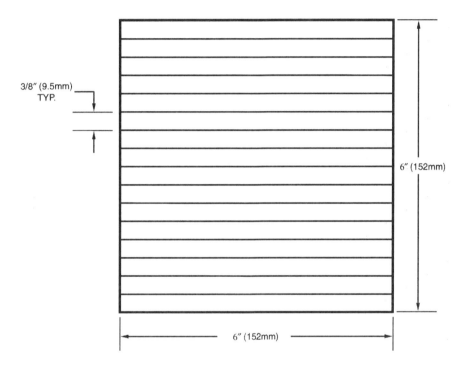

3. Answer the following questions. Set your machine and have the instructor check your settings.

   A. What type electrode covering does this electrode have?   3. A. _____

   B. With what type current can it be used?   B. _____

   C. In what position will you be welding?   C. _____

   D. What range of current should you use for this electrode?   D. _____
   (**Note:** Set the machine to the middle of this range.)

   E. What is the tensile strength of this electrode?   E. _____

4. Button your shirt or coveralls at the neck, put your welding helmet on and tighten the headband, put on your welding gloves, and prepare to weld.

5. Your metal should be placed on the table so that the 6″ (152mm) length is running from your left to right. See figure in #2.

6. Lower your arc helmet and attempt to strike the arc. Try both the straight up-and-down motion or the scratching motion. Once the arc is struck, carefully adjust the arc length to a distance approximately equal to the diameter of the electrode. Once the arc is stabilized, move forward in the left-to-right direction slowly for about 1″

(25.4mm) and withdraw the electrode so that the arc is stopped. (**Note**: If you are left-handed, you should weld in a right-to-left direction.)

Caution: Beginners sometimes cause the electrode to weld itself to the base metal when the arc is struck. If this occurs, proceed as follows:

- A. With your arc helmet still down, release the electrode from the electrode holder.
- B. Raise your arc helmet.
- C. Hang the electrode holder on the insulated hanger.
- D. Bend the electrode back and forth until it breaks away from the base metal.
- E. Reinstall the electrode in the electrode holder.
- F. Lower your arc helmet and try again.

7. If the electrode sticks a great deal, increase the current setting on the machine about 10 A and try again. If the electrode spatters a great deal, try decreasing your arc gap or decrease the current setting on the machine about 10 A.

8. Try to restrike the arc just ahead of the first attempt and run another 1″ (25.4mm) bead. Then stop the arc.

9. Continue striking and stopping the arc until two of the lines on your plate are covered with 1″ (25.4mm) long beads.

Instructor's initials: _____

Job 12A-3

LEARNING OBJECTIVE	
✶ In this job, you will be running (creating) several arc beads.	Name _____ Date _____ Class _____ Instructor _____ Grade _____

1. Use the same side of the plate you were given to practice striking the arc in the previous job. Obtain six 1/8" (3.2mm) diameter, E6012 electrodes to begin this job.

2. Set the machine polarity. Set the amperage at the mid-range for the electrode being used.

3. Lower your arc helmet. Strike the arc, and run a stringer bead (no sideways electrode motion) continuously along the marked line.

4. The bead should be about three times as wide as the electrode diameter. The ripple shape at the rear of the weld pool should be bullet-shaped if the forward speed is correct. The electrode should be changed when the total length of the electrode gets to be about 2 1/2" (63.5mm) or otherwise as specified by your instructor. Place a new electrode into the holder and restart the bead as described in Heading 12.5.4 of the text.

5. Wear chipping goggles whenever you are cleaning your beads. Chip and wire brush your bead until the slag coating is removed.

6. Check the appearance of your first bead against those shown in Figure 12-25 in the text. If the bead is not correct, change what needs to be changed before you run the second bead.

7. Run five additional beads, one marked line apart from each other.

8. Chip and clean each bead prior to beginning the next bead. Make corrections to the machine settings, your electrode angle, travel speed, arc gap, bead width, and bead shape as required after you inspect and evaluate each bead.

INSPECTION

The beads should be straight, about 3/8" (9.5mm) wide. They must have a uniform buildup, evenly spaced ripples, and good fusion with no overlap or undercutting at the toes.

Instructor's initials: _____

Job 12A-4

LEARNING OBJECTIVE

★ You will run several weaving beads using an E6012 electrode.

Name _____
Date _____ Class _____
Instructor _____
Grade _____

1. Obtain about six 1/8" (3.2mm) diameter, E6012 electrodes to begin this job.

2. Using a rule and a piece of soapstone or chalk, draw lines 3/8" (9.5mm) apart across the full 6" (152mm) length of the metal on both sides. See figure below.

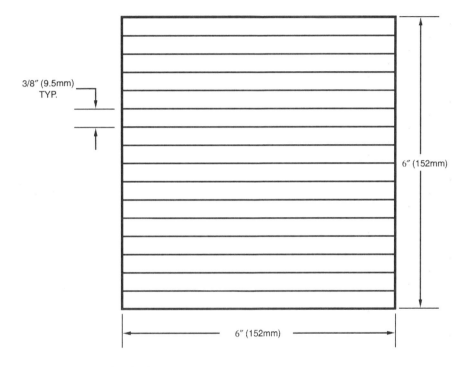

3. Set the machine amperage and polarity.

4. For this job, use the side opposite the one used for the previous two jobs.

5. The second line from the edge should be used as the center point of your bead. Make a weaving bead approximately 3/4" (19mm) wide or 6 times the electrode diameter.

6. Chip and wire brush this bead.

7. Read the bead. Inspect it, and decide what changes need to be made to improve its appearance and quality.

8. Use the 4th, 6th, and 8th lines as centers and make three additional weaving beads.

INSPECTION

The beads should be straight, about 3/4" (19mm) wide. They must have a uniform buildup, evenly spaced ripples, and good fusion with no overlap or undercutting at the toes.

Instructor's initials: _____

Job 12A-5

LEARNING OBJECTIVE
* You will run stringer beads with E6010, E6011, E6013, and other electrodes as selected by the instructor.

Name _____
Date _____ Class _____
Instructor _____
Grade _____

1. Using 1/8" (3.2mm) E6010 electrodes, make five stringer beads next to the weaving beads completed in Job 12A-4.

2. Answer the following questions before beginning to weld. (See Figures 12-17 and 12-19 of the text.)

   A. What amperage range is suggested? 2. A. _____

   B. What type of current is required? B. _____

   C. What type of covering is on this electrode? C. _____

3. Clean, inspect, and read each bead before making the next bead. Change any setting or other variable required to make the next bead better.

4. Using 5/32" (4mm) E6011 electrodes, make five weaving beads on the side opposite the weaving beads made in #1. Weld wherever space is available or in between or on top of a previous bead. Answer the following questions before beginning to weld:

   A. What amperage range is suggested? 4. A. _____

   B. What type of current may be used? B. _____

   C. What type of covering is used? C. _____

5. Run five stringer beads using the same electrode type and size.

6. Read each bead and make corrections before making the next bead.

7. Using 3/32" (2.4mm) E6013 electrodes, make five stringer beads next to each other. Weld wherever space is available or in between or on top of previous beads. Answer the following questions before starting to weld:

   A. What amperage range is suggested? 7. A. _____

   B. What type current may be used? B. _____

   C. What type of covering is on this electrode? C. _____

8. Clean and read each bead before making the next bead.

INSPECTION

All beads should be even in width, straight, with uniform buildup and ripple spacing. No overlap or undercutting should be present at the toes of the weld. The stringer bead width should be three times the electrode diameter. Weaving beads should be six times the electrode diameter.

Instructor's initials: _____

Lesson 12B

ARC WELDING IN THE FLAT WELDING POSITION

Name _____ Date _____ Lesson Grade _____
Class _____ Instructor _____

LEARNING OBJECTIVE
★ You will be able to do SMAW in the flat welding position.

INSTRUCTIONS
Carefully read Headings 12.6 through 12.7.5 of the text. Also study Figures 12-28 through 12-50 in the text. Then complete the following problems.

1. In ac, the _____ _____ direction of current virtually cancels the magnetic blow effect.

   1. _____

2. When using dc, the arc has a tendency to _____ as a weld nears the end of the joint.
   A. blow toward the end of the joint
   B. blow toward the beginning of the joint
   C. remain steady
   D. blow to the right side
   E. blow to the left side

   2. _____

3. What type of weld is described below?

   The weld axis is horizontal, and the weld face is horizontal or near-horizontal.

   3. _____
   _____

4. Name the parts of a fillet weld shown below.
   A. _____
   B. _____
   C. _____
   D. _____
   E. _____

Copyright Goodheart-Willcox Co., Inc.

145

5. Name the parts of the groove weld shown below.

   A. _____

   B. _____

   C. _____

   D. _____

   E. _____

   F. _____

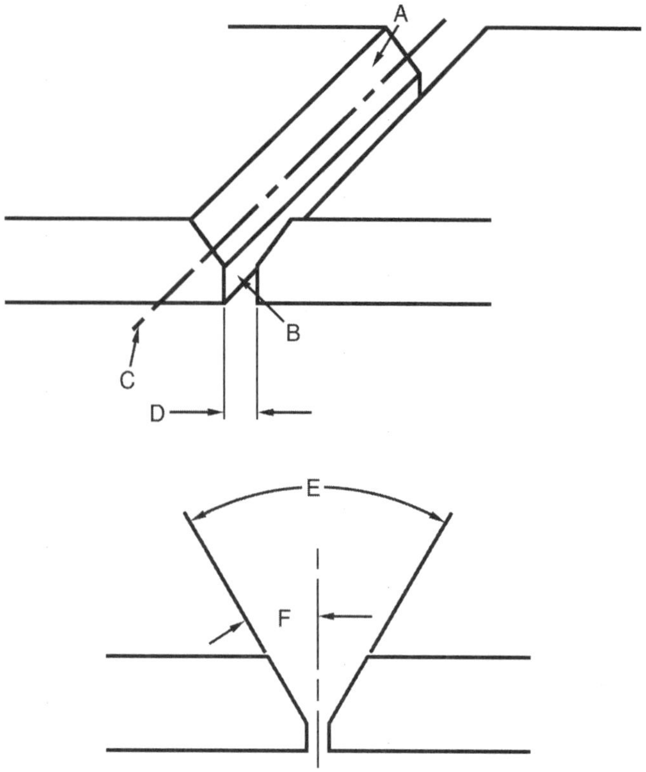

6. What type of defect is shown in the welds below?

   A. _____

   B. _____

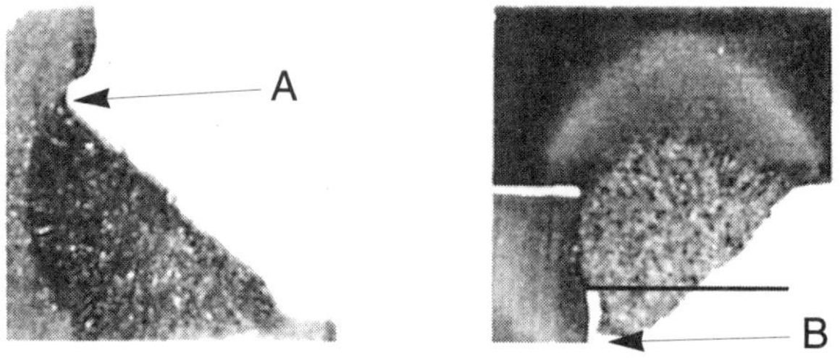

7. How wide should the bead generally be when making an edge weld? _____

   _____

   _____

8. Name the parts of a groove weld in progress shown below.

   A. _____

   B. _____

   C. _____

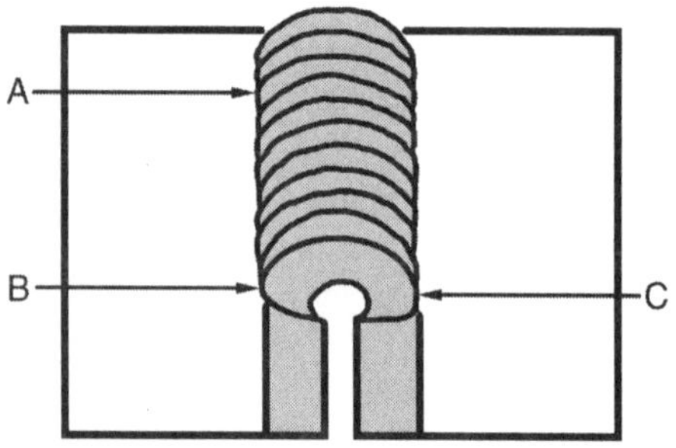

9. A. How many beads were used to make the groove weld below?   9. A. _____

   B. What is the proper name of the first bead?   B. _____

10. List three corrective measures that may be taken to prevent or reduce arc blow. _____

_____
_____
_____
_____
_____
_____

Job 12B-1

LEARNING OBJECTIVE

★ In this job, you will learn to make a square-groove weld on an edge joint in the flat welding position.

Name _____
Date _____ Class _____
Instructor _____
Grade _____

1. Obtain two pieces of mild steel that measure 1/4" × 5" × 5" (6.4mm × 127mm × 127mm).
2. Obtain five E6010 electrodes of the smallest suggested diameter.
3. Refer to Figures 12-17 and 12-19 in the text, and answer the questions below for the E6010 electrode used:

    A. What diameter should be used?
    B. What type and polarity or current should be used?
    C. What current range should be used?

    3. A. _____
    B. _____
    C. _____

4. What number welding filter lens should be used? (Refer to Figure 11-59 in the text.)

    4. _____

5. Place the pieces together and tack weld the weldment twice on each side.
6. Make an edge weld in the flat welding position on each of the four sides as shown in the drawing below.

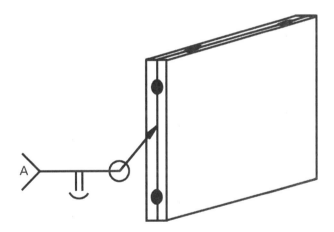

A—FLAT POSITION, WEAVING BEAD TO EDGES

INSPECTION

Your beads should be complete, with no voids or low spots where the beads stop and start. The beads should have evenly spaced ripples, with an even buildup. There should be no overlapping at the metal edges.

Instructor's initials: _____

Job 12B-2

LEARNING OBJECTIVE

* In this job, you will learn to produce a fillet weld on a lap joint in the flat welding position.

Name _____
Date _____ Class _____
Instructor _____
Grade _____

1. Obtain four pieces of mild steel that measure 1/4" × 1 1/2" × 5" (6.4mm × 38mm × 127mm).
2. Obtain five E6011 electrodes of the largest suggested diameter.
3. Refer to Figures 12-17 and 12-19 of the text and answer the questions below for the electrode and diameter used.

   A. What diameter electrode should be used? 3. A. _____
   B. What type and polarity or current should be used? B. _____
   C. What current range should be used? C. _____
4. What number welding lens filter should be used? (Refer to Figure 11-59 in the text.) 4. _____
5. Place the pieces together, and tack weld them twice on each side.
6. Make two fillet welds as shown in the drawing below.

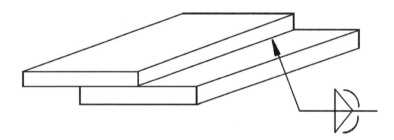

INSPECTION

Your fillet welds should be straight and even in width. The ripples in the beads should be evenly spaced, and the beads should be convex. There should be no undercutting or overlapping.

Instructor's initials: _____

Job 12B-3

LEARNING OBJECTIVE

* In this job, you will make a fillet weld on a T-joint in the flat welding position.

Name _____
Date _____ Class _____
Instructor _____
Grade _____

1. For this job, you will need two pieces of mild steel that measure 3/16" × 1 1/2" × 5" (4.8mm × 38mm × 127mm).

2. Obtain five E6012 electrodes of the proper size.

3. Refer to Figures 12-17 and 12-19 in the text, and answer the following questions.

    A. What diameter electrode should be used? 3. A. _____

    B. What type and polarity or current should be used? B. _____

    C. What current range should be used with this electrode? C. _____

4. What number filler lens should be used? 4. _____
   (Refer to Figure 11-59 in the text.)

5. Place the pieces into the proper position and tack weld them on each side of the joint about 3" (76mm) apart.

6. Make the welds as shown in the drawing below.

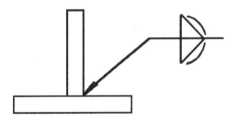

INSPECTION

Your fillet weld should be straight, with an even width, and evenly spaced ripples. They should be convex in shape. They should be free from undercutting or overlapping.

Instructor's initials: _____

Job 12B-4

LEARNING OBJECTIVE

★ In this job, you will learn to make a bevel-groove weld on a butt joint in the flat welding position.

Name _____
Date _____ Class _____
Instructor _____
Grade _____

1. Obtain four pieces of mild steel that measure 1/4″ × 1 1/2″ × 5″ (6.4mm × 38mm × 127mm).

2. You will need five E6013 electrodes for this job.

3. Refer to Figures 12-17 and 12-19 in the text, and answer the following questions:

    A. What diameter electrode should be used?   3. A. _____

    B. What type and polarity or current should be used?   B. _____

    C. What current range should be used?   C. _____

4. What number welding filter lens should be used?   4. _____

5. Prepare the bevel. Place the pieces in their proper position and tack weld each joint.

6. Make the welds as shown in the drawing below. Be certain that the welds are made on the correct side of the metal.

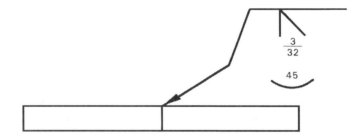

INSPECTION

Your welds should be straight, even in width, with all beads having evenly spaced ripples. There must be 100% penetration, with the penetration showing on the root side of each joint.

Instructor's initials: _____

Lesson 12C

ARC WELDING OUT OF POSITION

Name _____ Date _____ Lesson Grade _____
Class _____ Instructor _____

LEARNING OBJECTIVE
★ You will be able to do SMAW in the horizontal position.

INSTRUCTIONS
Carefully read Headings 12.7.5 through 12.8 of the text. Also study Figures 12-51 through 12-62 in the text. Then complete the following problems.

1. Why is the electrode pointed upward at about 20° when performing a horizontal weld? _____
   _____
   _____

2. Gravity causes the metal in the air to fall as it travels across the arc from the electrode to the base metal. What may be done to reduce this tendency while welding in the horizontal, vertical, and overhead positions?   2. _____
   A. Increase the electrode diameter
   B. Increase the welding current
   C. Increase the arc gap slightly
   D. Decrease the arc gap slightly
   E. Increase the welding speed

3. Circle the ways which may be successfully used to eliminate undercutting.
   A. Increase the welding current slightly.
   B. Decrease the welding current slightly.
   C. Stop momentarily at the end of each motion or swing of the electrode.
   D. Travel faster.
   E. Use a longer arc length.

4. When welding a lap joint, which piece of metal requires more heat? _____
   _____
   _____

5. When making a fillet weld in a lap or inside corner joint in the horizontal position, regardless of the motion shape used, it should have a _____ slant.

5. _____

6. Special _____ _____ are often used to ensure that all welds on a weldment can be done in the flat welding position.

6. _____

7. List five ways to prevent filler metal sagging. _____
_____
_____
_____
_____
_____
_____

8. What should a welder be aware of if a covered electrode is bent? _____
_____

9. Describe the horizontal welding position. (Refer to other areas of the text, if necessary, to find this answer.)
_____
_____

10. What type of helmet is the welder wearing in Figure 11-61? _____
_____

Job 12C-1

LEARNING OBJECTIVE

★ In this job, you will demonstrate your ability to weld a fillet weld on a lap joint in the horizontal welding position.

Name _____
Date _____ Class _____
Instructor _____
Grade _____

1. Obtain three pieces of mild steel measuring 1/4" × 1 1/2" × 5" (6.4mm × 38mm × 127mm) and six E6012 electrodes. (See #2A for the diameter.)

2. Answer the following questions before starting to weld. (Refer to Figure 12-19 in the text.)

   A. What is the smallest diameter electrode that is recommended for welding your metal?   2. A. _____

   B. What amperage range should be used?   B. _____

3. Using ac and the electrode diameter and amperage settings above, tack weld the weldment in the flat position.

4. After tack welding, place the weldment into a weld positioning fixture or prop it against firebricks and make the welds shown below. **Note**: The weldment may be turned as necessary so that all welds are made in the horizontal welding position.

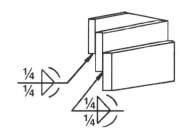

5. Read each weld prior to making the next. Make whatever changes may be necessary to improve the next weld made.

INSPECTION
All fillet welds should be even in width, with evenly spaced ripples, and with no undercutting or overlapping.

Instructor's initials: _____

Job 12C-2

LEARNING OBJECTIVE

★ In this job, you will weld fillet welds on a T-joint in the horizontal position.

Name _____
Date _____ Class _____
Instructor _____
Grade _____

1. Obtain three pieces of 1/4" × 1 1/2" × 5" (6.4mm × 38mm × 127mm) mild steel and six E6011 electrodes. (See #2a for the diameter.)

2. Answer the following questions before beginning to weld. (Refer to Figure 12-19 in the text.)

   A. What is the largest diameter electrode recommended for this metal thickness?
   2. A. _____

   B. What range of amperage should be used?
   B. _____

   C. Ac is not to be used! What dc polarity will be used? (See Figure 12-17 of the text.)
   C. _____

3. Use the dc polarity, electrode diameter, and the amperage at the midpoint of the suggested amperage range shown in #2 above. Tack weld the weldment shown below in the flat welding position.

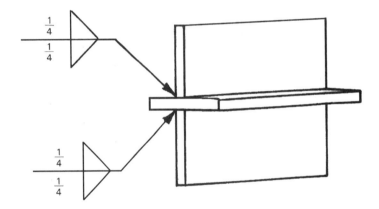

4. After tack welding, place the weldment in a weld positioning fixture or prop it against firebricks so that one joint is in the horizontal welding position. Complete all four welds. **Note:** The weldment may be turned as necessary so that all welds are made in the horizontal welding position.

5. Read each weld prior to making the next. Change whatever may need changing so that the next weld is of a higher quality.

INSPECTION

All fillet welds should have evenly spaced ripples, a uniform width, and have no undercutting or overlapping.

Instructor's initials: _____

Job 12C-3

LEARNING OBJECTIVE

★ In this job, you will be required to weld a bevel-groove butt joint and a V-groove outside corner joint in the horizontal welding position.

Name _____
Date _____ Class _____
Instructor _____
Grade _____

1. Obtain five pieces of mild steel measuring 1/4" × 1 1/2" × 5" (6.4mm × 38mm × 127mm) and six 1/8" E6012 electrodes.

2. Answer the following questions before beginning to weld:

   A. What amperage range is recommended? 2. A. _____

   B. Ac is not to be used! What polarity dc current is suggested? B. _____

3. Bevel the correct edge of two pieces of metal for use in the bevel groove butt joint shown in #6 below.

4. Tack weld all the pieces to create the weldment shown in #6 below. These tack welds should be made in the flat welding position.

5. After tack welding, place the weldment into a weld positioning fixture or lean it against firebricks.

Note: The weldment may be turned as required to make all welds in the horizontal welding position.

6. Complete all welds as required in the welding drawing below:

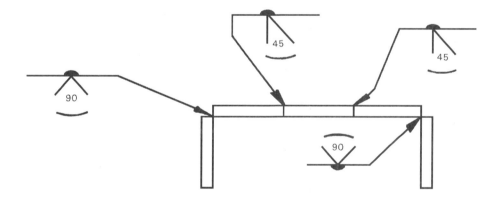

7. Read each weld before making the next one. Make required changes so that each weld is improved as required.

INSPECTION

All welds must have 100% penetration. The weld beads must have an even width, uniformly spaced ripples, be built-up, and be free of undercutting and overlapping.

Instructor's initials: _____

Lesson 12D

ARC WELDING OUT OF POSITION

Name _____ Date _____ Lesson Grade _____
Class _____ Instructor _____

LEARNING OBJECTIVE
★ You will be able to do SMAW in the vertical and overhead positions.

INSTRUCTIONS
Carefully read Headings 12.7.5 through 12.8 of the text. Also study Figures 12-51 through 12-62 in the text. Then complete the following problems.

1. A(n) _____ direction of vertical welding is preferred to reduce slag inclusions and produce the strongest welds.

    1. _____

2. Why is a "flip motion" used when performing welds in the horizontal, vertical, or overhead position? _____
_____
_____

3. *True or False?* When performing a vertical fillet weld on a lap joint, the forward motion and any flip motion is done on the surface, not on the part with the exposed edge.

    3. _____

4. Welds made in the _____ position are generally considered to be the most dangerous and the hardest to perform.

    4. _____

5. _____ must be worn for safety when welding in the overhead welding position.
    A. A hat or cap and an arc helmet
    B. Coveralls or workclothes without open pockets, buttoned at the collar
    C. Leather quality gloves
    D. A leather cape or coat
    E. All of the above.

    5. _____

6. What do the letters SMAW mean? _____
_____
_____

7. The welder in the figure below is using a welding fume _____ _____ and a powered _____ _____ as he welds a large assembly.

7. _____
   _____
   _____

8. Describe the steps and movements required to restart an arc and continue a bead.

   A. _____

   B. _____

   C. _____

   D. _____

9. *True or False?* Vertical welds should have the same appearance as a weld made in the flat welding position.

9. _____

10. When lap welding, when are the electrode and weld pool moved away from the edge to the surface of the metal? _____

Job 12D-1

LEARNING OBJECTIVE

★ In this job, you will use SMAW to make a fillet weld on a vertical lap joint.

Name _____
Date _____ Class _____
Instructor _____
Grade _____

Note: Weak vertical welds are produced when the molten slag is included (trapped) in the weld.

1. Obtain three pieces of 3/32" (2.4mm) mild steel that measure 1 1/2" × 5" (31.8mm × 127mm) and six E6012 electrodes.

2. Answer the following questions before beginning to weld.

   A. What diameter electrodes should be used? 2. A. _____

   B. What amperage range is suggested? B. _____

   C. What type current and polarity may be used? C. _____

Note: See Figures 12-17 and 12-19 in the text.

3. Tack weld each joint to form the weldment shown in #5 below. The tack welds may be made in the flat welding position.

4. After tacking, place the weldment in a weld positioning fixture or prop it against firebricks.

5. Each fillet weld shown in the drawing below shall be made in the vertical position.

Note: Use the current and polarity indicated on the welding symbol.

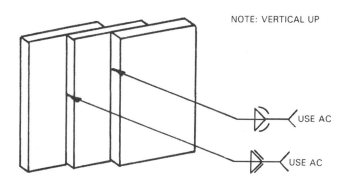

6. Read each weld prior to making the next weld. Make the changes required to improve the next weld made.

INSPECTION

Each fillet weld must be convex in shape, have an even width, and evenly spaced ripples. The weld should have no undercutting or overlapping.

Instructor's initials: _____

Job 12D-2

LEARNING OBJECTIVE

★ In this job, you will demonstrate your ability to perform fillet welds on T-joints in the vertical welding position.

Name _____
Date _____ Class _____
Instructor _____
Grade _____

1. Obtain four pieces of mild steel which measure 1/4" × 1 1/2" × 5" (6.4mm × 38mm × 127mm) and six E6013 electrodes.

2. Answer the following questions before beginning to weld:

    A. What is the smallest diameter electrode suggested? 2. A. _____

    B. What is the current range suggested in Figure 12-19 in the text? B. _____

    C. What type of current and polarity is suggested? C. _____

3. Tack weld two weldments as shown in #5 below. The tack welding may be done in the flat position.

4. Place the weldment into a weld positioning fixture or lean it against firebricks so the joint is in the vertical position.

5. Make the welds shown in the drawing below in the vertical position. **Note:** Make two of these weldments.

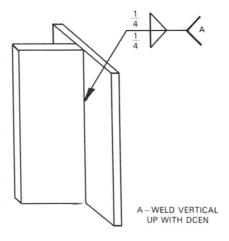

A—WELD VERTICAL UP WITH DCEN

INSPECTION

Each fillet weld should have an even width, smooth and even ripples, and be convex in shape. There should be no undercutting or overlapping.

Instructor's initials: _____

Job 12D-3

LEARNING OBJECTIVE

* In this job, you will learn how to perform a V-groove weld on a butt joint in the vertical welding position.

Name _____
Date _____ Class _____
Instructor _____
Grade _____

1. Obtain four pieces of mild steel that measure 1/4″ × 1 1/2″ × 5″ (6.4mm × 38mm × 127mm) and six E6010 electrodes.

2. Before you start to weld, answer the following questions:

   A. What is the smallest diameter electrode suggested in Figure 12-19 in the text?

   2. A. _____

   B. What current type and polarity is suggested in Figure 12-17 in the text?

   B. _____

   C. What is the suggested amperage range?

   C. _____

3. Wearing grinding goggles, grind, flame cut, or machine the edges of the metal as required in the weld symbol in #6.

4. Tack weld the metal to form the weldment shown in #6 below. Your tack welds may be made in the flat position.

5. After tack welding, place the weldment into a weld positioning fixture or lean it against firebricks.

6. Each weld must be made in the vertical welding position and according to the drawing below.

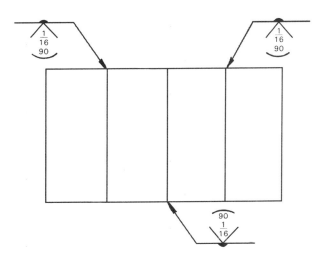

7. Read each weld before making the next weld. Make any corrections necessary to improve the next weld made.

INSPECTION

Each weld must have a 100% penetration showing on the root side. The weld beads must be even in width, have evenly spaced ripples, and have no undercutting or overlapping.

Instructor's initials: _____

Job 12D-4

LEARNING OBJECTIVE
* In this job, you will learn how to make a fillet weld in a lap joint in the overhead welding position.

Name _____
Date _____ Class _____
Instructor _____
Grade _____

1. Obtain three pieces of mild steel that measure 1/8" × 1 1/2" × 5" (3.2mm × 38mm × 127mm) and four E6011 electrodes.

2. Refer to Figures 12-17 and 12-19 in the text to answer the following questions:

   A. What is the smallest diameter electrode recommended? 2. A. _____

   B. What type current and polarity is recommended? B. _____

   C. What is the recommended current range? C. _____

3. Tack each joint to form the weldment shown in #6. Your tack welding may be done in the flat welding position.

4. After tack welding, place the weldment into a weld positioning fixture at a height above your head.

Caution: When welding in the overhead position, be certain to wear a cap, cape or leather jacket, gloves, and a helmet. Flash goggles may also be worn.

5. Each fillet weld must be made in the overhead position. Turn the weldment over for the last two welds.

Note: Use the type current shown on the weld symbol.

6. Make the welds shown in the drawing below in the overhead position. Use DCEP and not ac.

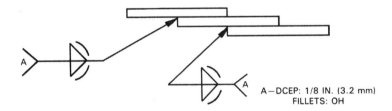

A—DCEP: 1/8 IN. (3.2 mm)
FILLETS: OH

7. Read each weld before making the next weld.

INSPECTION
Your fillet welds must have a convex face, even ripples, and even width. There should be no overlapping or undercutting visible.

Instructor's initials: _____

Job 12D-5

LEARNING OBJECTIVE

★ In this job, you will learn to make a fillet weld on a T-joint in the overhead welding position.

Name _____
Date _____ Class _____
Instructor _____
Grade _____

1. Obtain three pieces of mild steel that measure 1/4″ × 1 1/2″ × 5″ (6.4mm × 38mm × 127mm) and six E6012 electrodes.

2 Refer to Figures 12-17 and 12-19 in the text to answer the following questions:

   A. What is the smallest diameter electrode recommended?   2. A. _____

   B. What type current and polarity is recommended?   B. _____

   C. What is the recommended current range?   C. _____

3. Tack each joint to form the weldment shown in #6. The tack welding may be done in the flat position.

4. After tack welding, place the weldment into a weld positioning fixture and at a height above your head. **Caution:** When welding in the overhead welding position, be certain to wear a cap, cape or leather jacket, and leather spats to cover your shoe tops. Also, wear coveralls or workclothes buttoned at the collar, gloves, and a helmet. Flash goggles may also be worn.

5. Each fillet weld must be made in the overhead position. Turn the weldment over for the last two welds.

Note: Use the type current shown on the weld symbol.

6. Make the welds shown in the drawing below with ac in the overhead position.

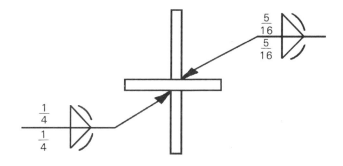

Note: It may be necessary to alter your travel speed or current settings to obtain a different size fillet on each side of the weldment.

7. Read each weld before making the next weld.

INSPECTION

Your fillet welds must have a convex face, even ripples, and even width. There should be no overlapping or undercutting visible.

Instructor's initials: _____

Job 12D-6

LEARNING OBJECTIVE
* In this job, you will learn to make a bevel-groove weld in a butt joint in the overhead welding position.

Name _____
Date _____ Class _____
Instructor _____
Grade _____

1. Obtain three pieces of mild steel that measure 1/4" × 1 1/2" × 5" (6.4mm × 38mm × 127mm) and six E6010 electrodes.

2. Refer to Figures 12-17 and 12-19 in the text and answer the following questions:

    A. What is the smallest diameter electrode recommended?   2. A. _____

    B. What type current and polarity is recommended?   B. _____

    C. What is the recommended current range?   C. _____

3. Tack each joint to form the weldment shown in #6. Your tack welding may be done in the flat position.

4. After tack welding, place the weldment into a weld positioning fixture and at a height above your head.

5. Each bevel groove weld must be made in the overhead welding position.

6. Make the weldment shown below in the overhead position.

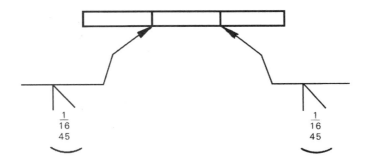

7. Read each weld before making the next weld.

INSPECTION
Your fillet welds must have a convex face, even ripples, and even width. There should be no overlapping or undercutting visible.

Instructor's initials: _____

Job 14A-1

LEARNING OBJECTIVE	Name _____
✱ In this job, you will learn to perform a fillet weld on a lap joint in the flat welding position. GTAW will be used.	Date _____ Class _____ Instructor _____ Grade _____

1. Obtain three pieces of 1/16″ (1.6mm) mild steel which measure 1 1/2″ × 5″ (38.1mm × 127mm).

2. Thoroughly clean oil, rust, and dirt from these pieces.

3. The following questions must be answered prior to starting to weld. Refer to Figure 14-19 in the text.

   A. What diameter electrode should be used? 3. A. _____

   B. What diameter filler rod will be used? B. _____

   C. What amperage range will be used? C. _____

   D. What type of shielding gas is used? D. _____

   E. The rate of gas flow is _____ cfh (_____ L/min). E. _____

   F. What type of current is to be used? F. _____

   G. What type of tungsten electrode is to be used? G. _____

   H. Is the electrode to be pointed or balled? H. _____

4. Try a #6-#8, 3/8″-1/2″ (9.5mm–12.7mm) diameter nozzle for this job. Install the torch nozzle and prepare the electrode. Place the electrode into the torch using the proper size collet and the proper extension for the electrode.

5. Set the machine for the proper current and polarity.

6. Using the correct shielding gas, start the gas flow. Set the correct flow rate on the flowmeter.

7. Tack weld the metal pieces to form the weldment shown below.

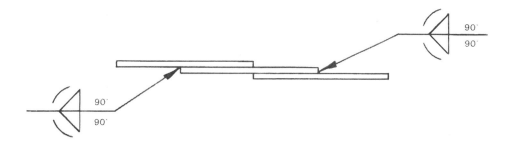

8. Install the weldment into a weld positioning fixture or lean it against firebricks to weld it in the flat position. Weld all the fillets as shown in the flat welding position. Turn the metal as required.

9. "Read" each weld and change whatever is required to improve the quality of the next weld.

INSPECTION

Your fillet welds should be convex, even in width, with evenly spaced ripples, and no overlap or undercut.

Instructor's initials: _____

Job 14A-2

LEARNING OBJECTIVE

* In this job, you will demonstrate your ability to weld a fillet weld on an inside corner joint in the flat welding position.

Name _____
Date _____ Class _____
Instructor _____
Grade _____

1. Obtain the following mild steel pieces.

    A. Two pieces 1/8" × 1 1/2" × 5" (3.2mm × 38mm × 127mm).

    B. One piece 1/8" × 3" × 5" (3.2mm × 76.2mm × 127mm).

2. Thoroughly clean all pieces.

3. The following questions must be answered before you start to weld. Refer to Figure 14-19 in the text.

    A. What diameter electrode should be used?   3. A. _____
    B. What diameter filler rod will be used?   B. _____
    C. What amperage range will be used?   C. _____
    D. What type of shielding gas is used?   D. _____
    E. The rate of gas flow is _____ cfh (_____ L/min).   E. _____
    F. What type of current is to be used?   F. _____
    G. What type of tungsten electrode is to be used?   G. _____
    H. Is the electrode to be pointed or balled?   H. _____

4. Try a #8 – #10 torch nozzle for this job. Install the torch nozzle and prepare the electrode. Place the electrode into the torch using the proper collet diameter and electrode extension.

5. Set the machine for the proper current and polarity.

6. Using the correct shielding gas, start the gas flow. Set the flow rate on the flowmeter.

7. Tack weld the pieces to form the weldment shown below.

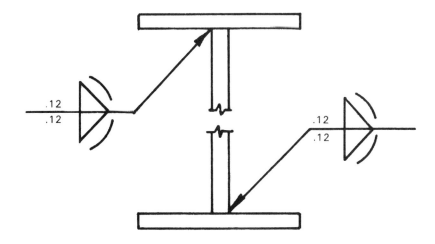

8. After tack welding, place the weldment into a weld positioning fixture or lean it against firebricks.
9. Weld all fillets in the flat welding position. Turn the metal as required to weld in the flat welding position.
10. "Read" each weld and change whatever is required to improve the quality of the next weld.

> **INSPECTION**
> Your fillet welds should be convex, even in width, with evenly spaced ripples and no overlap or undercut.
>
> Instructor's initials: _____

Job 14A-3

LEARNING OBJECTIVE

★ In this job, you will learn to weld a square groove weld on a butt joint.

Name _____
Date _____ Class _____
Instructor _____
Grade _____

1. Obtain five pieces of mild steel that measure 1/8″ × 1 1/2″ × 5″ (3.2mm × 38mm × 127mm).
2. Thoroughly clean the pieces.
3. The following questions must be answered prior to starting to weld. Refer to Figure 14-19 in the text.

    A. What diameter electrode should be used?   3. A. _____
    B. What diameter filler rod will be used?   B. _____
    C. What amperage range will be used?   C. _____
    D. What type of shielding gas is used?   D. _____
    E. The rate of gas flow is _____ cfh (_____ L/min). E. _____
    F. What type of current is to be used?   F. _____
    G. What type of tungsten electrode is to be used? G. _____
    H. Is the electrode to be pointed or balled?   H. _____

4. Try a #8 – #10 diameter nozzle for this job. Install the torch nozzle and prepare the electrode.
5. Set the machine for the proper current and polarity.
6. Using the correct shielding gas, start the gas flow. Set the correct flow rate on the flowmeter.
7. Tack weld the metal pieces to form the weldment shown below.

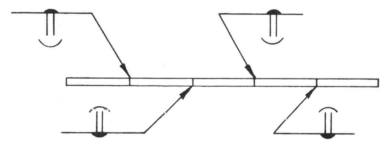

8. Install the weldment into a weld positioning fixture or place it on firebricks to weld it in the flat welding position.
9. Weld all the square groove butt welds shown in the flat welding position.
10. "Read" each weld and change whatever is required to improve the quality of the next weld.

INSPECTION

Your welds should have a convex face, with even width and evenly spaced ripples in the bead. All welds should show 100% penetration on the root side of the joint.

Instructor's initials: _____

Lesson 14B

GAS TUNGSTEN ARC WELDING

Name _____ Date _____ Lesson Grade _____
Class _____ Instructor _____

> **LEARNING OBJECTIVE**
>
> ✶ You will be able to prepare a weld. You will also be able to perform various welds in all positions.
>
> **INSTRUCTIONS**
>
> Carefully read Headings 14.4 through 14.15 of the text. Also study Figures 14-31 through 14-65 in the text. Then, complete the following problems.

1. Why shouldn't steel welding rods be copper coated for GTAW? _____
   _____
   _____
   _____

2. Name the items shown below.
   A. _____
   B. _____
   C. _____
   D. _____
   E. _____

3. Name the method of starting the arc that is shown above.   3. _____

4. In the drawing below, identify the angles and distances used in GTAW.

   A. _____

   B. _____

   C. _____

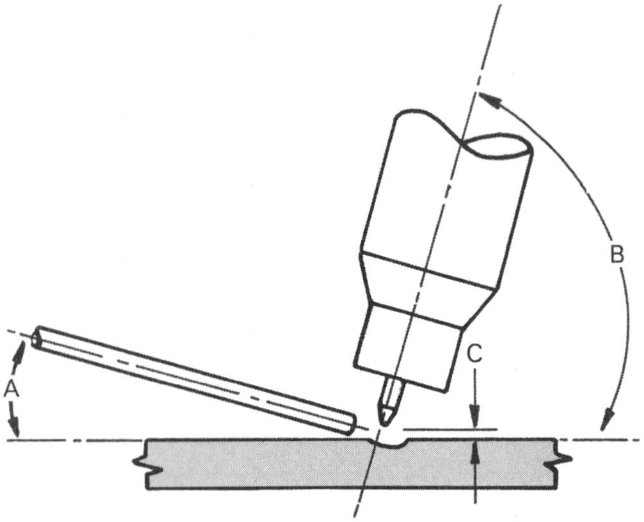

5. Why does the shielding gas automatically continue to flow for a short time after the GTAW arc is broken (stopped)? _____

_____

_____

_____

6. GTAW is not normally used on carbon steel above _____.   6. _____
   A. 1/8" (3.2mm)
   B. 1/4" (6.4mm)
   C. 3/8" (9.5mm)
   D. 1/2" (12.7mm)
   E. 5/8" (15.9mm)

7. What may happen if the flowmeter is not closed completely when the GTAW station is shut down? _____

_____

_____

_____

8. In all types of GTA welds, when the filler rod is added to the front   8. _____
   of the weld pool, the torch is _____.
   A. not moved until after the rod is removed
   B. moved toward the front of the crater
   C. moved toward the back of the crater
   D. moved to one side of the crater
   E. All of the above.

9. Identify the angles and dimensions as indicated for the horizontal fillet weld shown below.

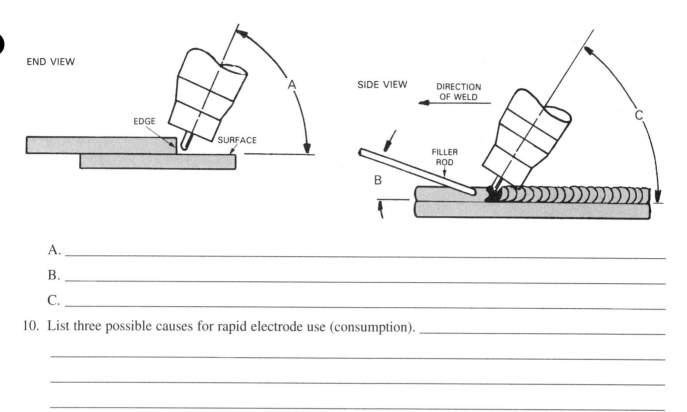

A. _____

B. _____

C. _____

10. List three possible causes for rapid electrode use (consumption). _____

_____

_____

_____

Job 14B-1

LEARNING OBJECTIVE

★ In this job, you will learn to produce fillet welds on a lap joint in the horizontal welding position using the GTAW process.

Name _____
Date _____ Class _____
Instructor _____
Grade _____

1. Obtain three pieces of mild steel that measure 1/8" × 1 1/2" × 5" (3.2mm × 38mm × 127mm).
2. Thoroughly clean oil, rust, and dirt from these pieces.
3. The following questions must be answered prior to starting to weld. Refer to Figure 14-19 in the text.

   A. What diameter electrode should be used?   4. A. _____
   B. What diameter filler rod will be used?   B. _____
   C. What amperage range will be used?   C. _____
   D. What type of shielding gas is used?   D. _____
   E. The rate of gas flow is _____ cfh (_____ L/min).   E. _____
   F. What type of current is to be used?   F. _____
   G. What type of tungsten electrode is to be used?   G. _____
   H. Is the electrode to be pointed or balled?   H. _____

4. Try a #8 – #10 diameter nozzle for this job. Install the torch nozzle and prepare the electrode. Place the electrode into the torch using the proper size collet and the proper extension for the electrode.
5. Set the machine for the proper current and polarity.
6. Using the correct shielding gas, start the gas flow. Set the correct flow rate on the flowmeter.
7. Tack weld the metal pieces to form the weldment shown below.

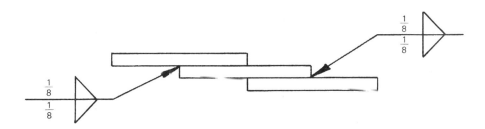

8. Install your weldment into a weld positioning fixture or support it with firebricks to weld it in the horizontal welding position. Weld all the fillets shown in the horizontal welding position.
9. "Read" each weld and change whatever is required to improve the quality of the next weld.

INSPECTION

Your fillet welds should be convex, even in width, with evenly spaced ripples and no overlap or undercut.

Instructor's initials: _____

Job 14B-2

LEARNING OBJECTIVE

★ In this job, you will learn to make a fillet weld on an inside corner and T-joint in the horizontal welding position.

Name _____
Date _____ Class _____
Instructor _____
Grade _____

1. Obtain mild steel with the following sizes:
   Four pieces: 1/16" × 1 1/2" × 5" (1.6mm × 38mm × 127mm).
   One piece: 1/16" × 5" × 6" (1.6mm × 127mm × 152mm).

2. Thoroughly clean oil, rust, and dirt from these pieces.

3. The following questions must be answered prior to starting to weld. Refer to Figure 14-19 in the text.

   A. What diameter electrode should be used?   3. A. _____
   B. What diameter filler rod will be used?   B. _____
   C. What amperage range will be used?   C. _____
   D. What type of shielding gas is used?   D. _____
   E. The rate of gas flow is _____ cfh (_____ L/min).   E. _____
   F. What type of current is to be used?   F. _____
   G. What type of tungsten electrode is to be used?   G. _____
   H. Is the electrode to be pointed or balled?   H. _____

4. Try a #6 – #8 diameter nozzle for this job. Install the torch nozzle and prepare the electrode. Place the electrode into the torch using the proper size collet and the proper extension for the electrode.

5. Set the machine for the proper current and polarity.

6. Using the correct shielding gas, start the gas flow. Set the correct flow rate on the flowmeter.

7. Tack weld the metal pieces to form the weldment shown below.

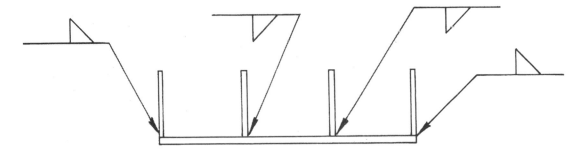

8. Install your weldment into a weld positioning fixture or support it with firebricks to weld it in the horizontal welding position. Weld all the fillets shown in the horizontal welding position.

9. "Read" each weld and change whatever is required to improve the quality.

INSPECTION

Your fillet welds should be convex, even in width, with evenly spaced ripples and no overlap or undercut.

Instructor's initials: _____

Job 14B-3

LEARNING OBJECTIVE

★ In this job, you will demonstrate your ability to make a square groove butt weld in the horizontal welding position.

Name _____
Date _____ Class _____
Instructor _____
Grade _____

1. Obtain five pieces of mild steel which measure 1/8″ × 1 1/2″ × 5″ (3.2mm × 38mm × 127mm).
2. Thoroughly clean these pieces.
3. The following questions must be answered prior to starting to weld. Refer to Figure 14-19 in the text.

   A. What diameter electrode should be used?   3. A. _____

   B. What diameter filler rod will be used?   B. _____

   C. What amperage range will be used?   C. _____

   D. What type of shielding gas is used?   D. _____

   E. The rate of gas flow is _____ cfh (_____ L/min).   E. _____

   F. What type of current is to be used?   F. _____

   G. What type of tungsten electrode is to be used?   G. _____

   H. Is the electrode to be pointed or balled?   H. _____

4. Try a #8 – #10 diameter nozzle for this job. Install the torch nozzle and prepare the electrode. Place the electrode into the torch using the proper size collet and the proper extension for the electrode.
5. Set the machine for the proper current and polarity.
6. Using the correct shielding gas, start the gas flow. Set the correct flow rate on the flowmeter.
7. Tack weld the metal pieces to form the weldment shown below.

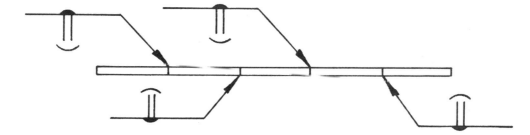

8. Install the weldment into a weld positioning fixture or lean it against firebricks to weld it in the horizontal welding position. Weld all square groove butt welds in the horizontal welding position.
9. "Read" each weld and change whatever is required to improve the quality of the next weld.

INSPECTION

Your welds should have a convex face, with an even width, and evenly spaced ripples in the bead. All welds should show 100% penetration on the root side of the joint.

Instructor's initials: _____

Job 14B-4

> **LEARNING OBJECTIVE**
>
> ✶ In this job, you will learn to perform a fillet weld on a lap joint in the vertical welding position.

Name _____
Date _____ Class _____
Instructor _____
Grade _____

1. Obtain three pieces of mild steel that measure 1/8" × 1 1/2" × 5" (3.2mm × 38mm × 127mm).
2. Thoroughly clean these pieces.
3. The following questions must be answered prior to starting to weld. Refer to Figure 14-19 in the text.

    A. What diameter electrode should be used?   3. A. _____
    B. What diameter filler rod will be used?   B. _____
    C. What amperage range will be used?   C. _____
    D. What type of shielding gas is used?   D. _____
    E. The rate of gas flow is _____ cfh (_____ L/min).   E. _____
    F. What type of current is to be used?   F. _____
    G. What type of tungsten electrode is to be used?   G. _____
    H. Is the electrode to be pointed or balled?   H. _____

4. Try a #8 – #10 diameter nozzle for this job. Install the torch nozzle and prepare the electrode. Place the electrode into the torch using the proper size collet and the proper extension for the electrode.
5. Set the machine for the proper current and polarity. Use a pulsed arc if the machine is equipped to furnish a pulsed arc.
6. Using the correct shielding gas, start the gas flow. Set the correct flow rate on the flowmeter.
7. Tack weld the pieces to form the weldment shown below.

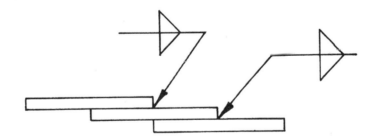

8. Install the weldment into a weld positioning fixture or lean it against firebricks so that all welds are performed in the vertical welding position. Weld all fillet welds in the vertical welding position.
9. "Read" each weld and change whatever is required to improve the quality of the next weld.

> **INSPECTION**
>
> Your fillet welds should be convex, even in width, with evenly spaced ripples and no overlap or undercut.
>
> Instructor's initials: _____

Job 14B-5

LEARNING OBJECTIVE	Name _____
★ In this job, you will learn to perform a fillet weld on a T-joint in the vertical welding position.	Date _____ Class _____ Instructor _____ Grade _____

1. Obtain three pieces of mild steel which measure 1/16″ × 1 1/2″ × 5″ (1.6mm × 38mm × 127mm).

2. Thoroughly clean these pieces.

3. The following questions must be answered prior to starting to weld. Refer to Figure 14-19 in the text.

   A. What diameter electrode should be used? 3. A. _____

   B. What diameter filler rod will be used? B. _____

   C. What amperage range will be used? C. _____

   D. What type of shielding gas is used? D. _____

   E. The rate of gas flow is _____ cfh (_____ L/min). E. _____

   F. What type of current is to be used? F. _____

   G. What type of tungsten electrode is to be used? G. _____

   H. Is the electrode to be pointed or balled? H. _____

4. Try a #6 – #8 diameter nozzle for this job. Install the torch nozzle and prepare the electrode. Place the electrode into the torch using the proper size collet and the proper extension for the electrode.

5. Set the machine for the proper current and polarity. Use a pulsed arc if the machine is equipped to furnish a pulsed arc.

6. Using the correct shielding gas, start the gas flow. Set the correct flow rate on the flowmeter.

7. Tack weld the metal pieces to form the weldment shown below.

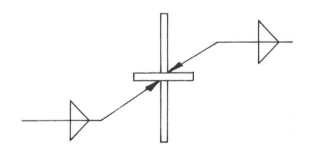

8. Install the weldment into a weld positioning fixture or lean it against firebricks so that all welds are performed in the vertical welding position. Weld two of the above weldments in the vertical welding position.

9. "Read" each weld and change whatever is required to improve the quality of the next weld.

INSPECTION

Your fillet welds should be convex, even in width, with evenly spaced ripples and no overlap or undercut.

Instructor's initials: _____

Job 14B-6

LEARNING OBJECTIVE

★ In this job, you will learn to perform a square-groove butt weld in the vertical welding position.

Name _____
Date _____ Class _____
Instructor _____
Grade _____

1. Obtain five pieces of mild steel which measure 1/8" × 1 1/2" × 5" (3.2mm × 38mm × 127mm).

2. Thoroughly clean these pieces.

3. The following questions must be answered prior to starting to weld. Refer to Figure 14-19 in the text.

   A. What diameter electrode should be used?   3. A. _____
   B. What diameter filler rod will be used?   B. _____
   C. What amperage range will be used?   C. _____
   D. What type of shielding gas is used?   D. _____
   E. The rate of gas flow is _____ cfh (_____ L/min).   E. _____
   F. What type of current is to be used?   F. _____
   G. What type of tungsten electrode is to be used?   G. _____
   H. Is the electrode to be pointed or balled?   H. _____

4. Try a #8 – #10 diameter nozzle for this job. Install the torch nozzle and prepare the electrode. Place the electrode into the torch using the proper size collet and the proper extension for the electrode.

5. Set the machine for the proper current and polarity. Use a pulsed arc if the machine is equipped to furnish a pulsed arc.

6. Using the correct shielding gas, start the gas flow. Set the correct flow rate on the flowmeter.

7. Tack weld the metal pieces to form the weldment shown below.

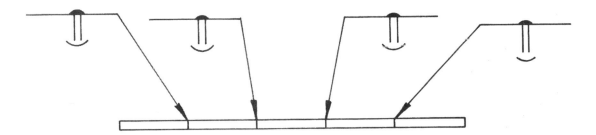

8. Install the weldment into a weld positioning fixture or lean it against firebricks so that all welds are performed in the vertical welding position. Weld all welds in the vertical welding position.

9. "Read" each weld and change whatever is required to improve the quality of the next weld.

INSPECTION

Your welds should have a convex face, with an even width, and evenly spaced ripples in the bead. All welds should show 100% penetration on the root side of the joint.

Instructor's initials: _____

Job 14B-7

LEARNING OBJECTIVE	
✱ In this job, you will learn to perform a fillet weld on a lap joint in the overhead welding position.	Name _____ Date _____ Class _____ Instructor _____ Grade _____

1. Obtain three pieces of 1/16" (1.6mm) mild steel that measure 1 1/2" × 5" (38mm × 127mm).

2. Thoroughly clean oil, rust, and dirt from these pieces.

3. The following questions must be answered before starting to weld. Refer to Figure 14-19 in the text.

    A. What diameter electrode should be used?   3. A. _____

    B. What diameter filler rod will be used?   B. _____

    C. What amperage range will be used?   C. _____

    D. What type of shielding gas is used?   D. _____

    E. The rate of gas flow is _____ cfh (_____ L/min).   E. _____

    F. What type of current is to be used?   F. _____

    G. What type of tungsten electrode is to be used?   G. _____

    H. Is the electrode to be pointed or balled?   H. _____

4. Try a #6 – #8 (3/8" or 9.5mm – 12.7mm) diameter nozzle for this job. Install the torch nozzle and prepare the electrode. Place the electrode into the torch using the proper size collet and the proper extension for the electrode.

5. Set the machine for the proper current and polarity. Use a pulsed arc if the machine is equipped to furnish a pulsed arc.

6. Using the correct shielding gas, start the gas flow. Set the correct flow rate on the flowmeter.

7. Tack weld the metal pieces to form the weldment shown below.

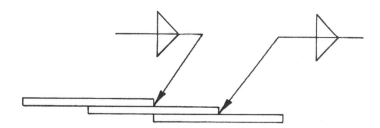

8. Install the weldment into a weld positioning fixture so that all welds are performed in the overhead welding position. Weld all fillet welds in the overhead welding position. Turn the metal as required.

9. "Read" each weld and change whatever is required to improve the quality of the next weld.

INSPECTION
Your fillet welds should be convex, even in width, with evenly spaced ripples and no overlap or undercut.   Instructor's initials: _____

Job 14B-8

LEARNING OBJECTIVE

* In this job, you will demonstrate your ability to perform a fillet weld on a T-joint in the overhead welding position.

Name _____
Date _____ Class _____
Instructor _____
Grade _____

1. Obtain three pieces of mild steel that measure 1/8" × 1 1/2" × 5" (3.2mm × 38mm × 127mm).
2. Thoroughly clean oil, rust, and dirt from these pieces.
3. The following questions must be answered prior to starting to weld. Refer to Figure 14-19 in the text.

   A. What diameter electrode should be used?
   B. What diameter filler rod will be used?
   C. What amperage range will be used?
   D. What type of shielding gas is used?
   E. The rate of gas flow is _____ cfh (_____ L/min).
   F. What type of current is to be used?
   G. What type of tungsten electrode is to be used?
   H. Is the electrode to be pointed or balled?

   3. A. _____
      B. _____
      C. _____
      D. _____
      E. _____
      F. _____
      G. _____
      H. _____

4. Try a #8 – #10 diameter nozzle for this job. Install the torch nozzle and prepare the electrode. Place the electrode into the torch using the proper size collet and the proper extension for the electrode.
5. Set the machine for the proper current and polarity. Use a pulsed arc if the machine is equipped to furnish a pulsed arc.
6. Using the correct shielding gas, start the gas flow. Set the correct flow rate on the flowmeter.
7. Tack weld the metal pieces to form the weldment shown below.

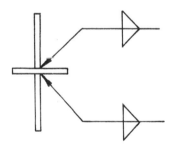

8. Install the weldment into a weld positioning fixture so that all welds are performed in the overhead welding position. Weld all fillets as shown in the overhead welding position. Turn the metal as required.
9. "Read" each weld and change whatever is required to improve the quality of the next weld.

INSPECTION

Your fillet welds should be convex, even in width, with evenly spaced tipples and no overlap or undercut.

Instructor's initials: _____

Job 14B-9

LEARNING OBJECTIVE ★ In this job, you will learn to perform the square groove butt weld in the overhead welding position.	Name _____ Date _____ Class _____ Instructor _____ Grade _____

1. Obtain five pieces of mild steel that measure 1/8″ × 1 1/2″ × 5″ (3.2mm × 38mm × 127mm).

2. Thoroughly clean oil, rust, and dirt from these pieces.

3. The following questions must be answered prior to starting to weld. Refer to Figure 14-19 in the text.

   A. What diameter electrode should be used?   3. A. _____

   B. What diameter filler rod will be used?   B. _____

   C. What amperage range will be used?   C. _____

   D. What type of shielding gas is used?   D. _____

   E. The rate of gas flow is _____ cfh (_____ L/min).   E. _____

   F. What type of current is to be used?   F. _____

   G. What type of tungsten electrode is to be used?   G. _____

   H. Is the electrode to be pointed or balled?   H. _____

4. Try a #8 – #10 diameter nozzle for this job. Install the torch nozzle and prepare the electrode. Place the electrode into the torch using the proper size collet and the proper extension for the electrode.

5. Set the machine for the proper current and polarity. Use a pulsed arc if the machine is equipped to furnish a pulsed arc.

6. Using the correct shielding gas, start the gas flow. Set the correct flow rate on the flowmeter.

7. Tack weld the metal pieces to form the weldment shown below.

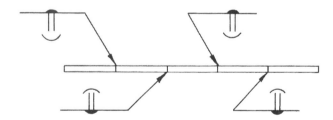

8. Install the weldment into a weld positioning fixture so that all welds are performed in the overhead welding position. Weld all welds as shown in the overhead welding position.

9. "Read" each weld and change whatever is required to improve the quality of the next weld.

INSPECTION

Your welds should have a convex face, with an even width, and evenly spaced ripples in the bead. All welds should show 100% penetration on the root side of the joint.

Instructor's initials: _____

Job 14B-10

LEARNING OBJECTIVE

★ In this job, you will learn to perform a groove weld on stainless steel in the 1G (flat) and 2G (horizontal) welding positions using the GTAW process.

Name _____
Date _____ Class _____
Instructor _____
Grade _____

1. Obtain eight pieces of stainless steel that measure 1/4″ × 1 1/2″ × 5″ (6.4mm × 38mm × 127mm).

2. Before starting to weld, answer the following questions regarding the variables that should be used.

   A. What is the best polarity for use with GTAW on stainless steel? 2. A. _____
   B. What diameter tungsten electrode is suggested? B. _____
   C. What diameter stainless filler wire is suggested? C. _____
   D. What amperage range will be used? D. _____
   E. What is the suggested shielding gas used on stainless steel? E. _____
   F. What gas flow rate should be set on the flowmeter? F. _____

3. Grind, flame cut, arc cut, or machine the edges of the eight pieces to make the joints shown in the figure below.

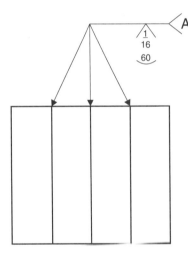

A— One in 1G position;
One in 2G position

4. Tack weld four pieces as shown in #3 above.

5. Place these pieces in a weld positioner and weld in the 1G position. "Read" each weld and make any required changes before making the next weld.

6. Tack weld the remaining four pieces as shown in the drawing in #3.

7. Place these pieces in a weld positioner and weld in the 2G position. "Read" each weld and make any required changes before making the next weld.

INSPECTION

Each weld must have 100% penetration shown on the root side. The weld beads must be even in width, have smooth evenly spaced ripples, and no overlapping or undercutting.

Instructor's initials: _____

Lesson 15A

GMAW EQUIPMENT AND SUPPLIES

Name _Ryan Norg_ Date _____ Lesson Grade _____
Class _Welding_ Instructor _____

LEARNING OBJECTIVE
✷ You will be able to describe the principles of GMAW.

INSTRUCTIONS
Carefully read the introduction to Chapter 15 and Headings 15.1. through 15.1.4 of the text. Also study Figures 15-1 through 15-7 in the text. Then complete the following problems.

1. Identify *all* of the statements below that apply to GMAW.   1. _BCE_
    A. It is also known as TIG
    B. It can be used in all positions
    C. It can be used to weld all commercially important metals
    D. It can be used with solid wire electrodes only
    E. It can be used to weld magnesium

2. What development made the use of GTAW more practical (useful)? _Development of constant voltage machines_

3. *True or False?* GTAW is done with DCEP (DCRP) only, except when using one kind of electrode.   3. _True_

4. When SMAW, the percentage of electrode metal actually deposited in the weld is 60%–70%. The percentage of electrode metal deposited when GMAW with a flux cored electrode is ____% to ____%.   4. _82% 92%_

5. List four ways in which metal may be transferred in the GMAW process. _Short Circuit_
Globular Spray Transfer Pulse Spray

6. Which of the GMAW metal transfer methods uses the lowest welding current? _Short circuit_

7. Which of the GMAW metal transfer methods uses the highest welding current? _Spray_

8. What GMAW metal transfer method is shown below? _Spray_

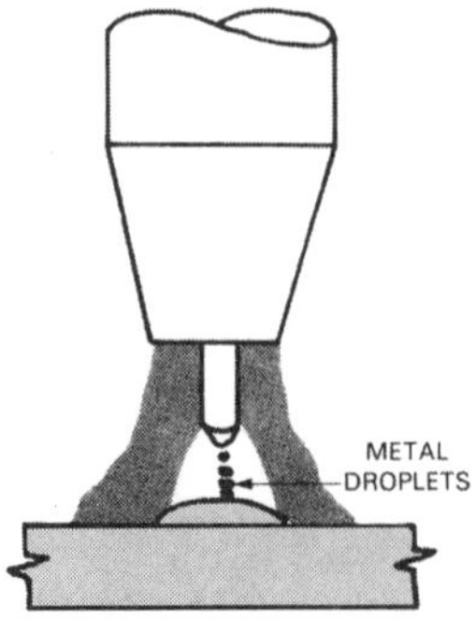

METAL DROPLETS

9. Circle the advantages of GMAW with the short circuiting method.
   A. Used well on thick metal sections
   B. There is no metal transfer through the arc
   C. Works well in all welding positions
   D. Works well on thin metal
   E. All of the above.

10. In short circuiting GMAW, when the electrode short circuits, the current may rise from 150 A to 500 A. What is used to slow down the possible rapid rise of the welding current? _Induction_

11. When using the globular transfer method, _____ causes the metal droplet to leave the electrode wire.
    A. magnetism
    B. high currents
    C. high voltages
    D. short arc
    E. the weight of the droplet

    11. _E_

12. The globular transfer method and a short arc length will cause a buried or _____ arc to occur.

    12. _Submerged_

13. The spray transfer method will deposit metal at _____ lbs./hr. as compared to _____ lbs./hr. with the short circuiting method.

    13. _6-12_
        _2-6_

14. Assume a welder earns $10.00/hour and a weldment requires 60 lbs. of metal.

Note: Use the highest lbs./hr. figure given by the AWS for both methods. Show your math below:
    A. What would be the cost of the welder when using the short circuiting method?
    B. What would be the cost of the welder when using the spray transfer method?

    14. A. _100 dollar_
        B. _50 dollars_

15. *True or False?* All spray transfer methods occur above the transition current. The transition current varies with the electrode diameter, the electrode composition, and the electrode extension.

16. The approximate transition current for a 0.045" diameter aluminum electrode is _____ amperes.

17. The pulsed spray transfer method uses a peak current for welding. The background current is used to _____ the arc.

18. Circle the advantages of the pulsed spray transfer method of GMAW.
    A. Higher current levels make it possible to weld out of position
    B. Thin metal sections can be welded
    C. This method creates very little spatter
    D. Larger diameter electrodes may be used
    E. Larger diameter electrodes are cheaper

19. *True or False?* The pulsed spray transfer method can be used in auto repair shops to weld light steel parts with very low heat inputs.

20. The spray transfer method of GMAW will occur only in a situation where the shielding gas is at least _____% argon.

15. True

16. 135 amp

17. maintain

19. True

20. 90%

Lesson 15B

GMAW EQUIPMENT AND SUPPLIES

Name _____ Date _____ Lesson Grade _____
Class _____ Instructor _____

LEARNING OBJECTIVE
★ You will be able to set up the GMAW station. You will also be able to select the electrode and shielding gas(es) and set the flow rate.

INSTRUCTIONS
Carefully read Headings 15.2 through 15.7 of the text. Then study Figures 15-8 and 15-40 in the text. Then complete the following problems.

1. When GMAW, changing the wire feed speed also changes the _____.   1. _____

2. When pulsed arc transfer is used in GMAW, the welding machine must be capable of setting the following:

    A. _____   2. A. _____

    B. _____ current   B. _____

    C. _____ current   C. _____

    D. _____ frequency if it is variable   D. _____

    E. _____ speed   E. _____

3. Name the controls indicated in the figure below:

    A. _____

    B. _____

    C. _____

    D. _____

    E. _____

Copyright Goodheart-Willcox Co., Inc.

4. To GMAW mild steel with the spray arc transfer method using a 0.030″ diameter electrode, what would the following values be?

   A. Arc voltage range

   B. Amperage range

   4. A. _____

   B. _____

5. To GMAW aluminum using the spray transfer method and using 1/16″ diameter electrode, what would the following values be?

   A. Arc voltage range

   B. Amperage range

   5. A. _____

   B. _____

6. What is the suggested amperage range for use with the spray transfer method, using a 3/32″ diameter electrode on aluminum?

   6. _____

7. Name the parts of the two-drive-roll GMAW wire drive shown below:

   A. _____

   B. _____

   C. _____

   D. _____

   E. _____

   F. _____

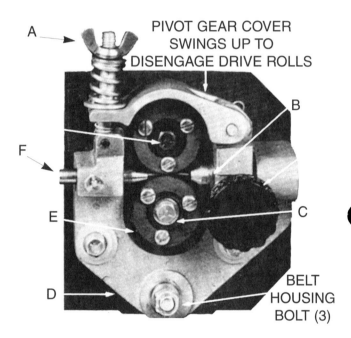

8. The wire rolls shown below are misaligned. Explain what would be moved, and in what direction, to realign these rolls. _____

_____

_____

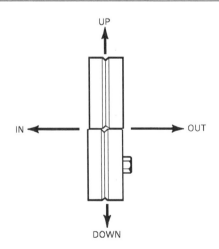

9. The rolls shown below are too far apart. What must be moved, and in what direction, to tighten the rolls?

_____

_____

_____

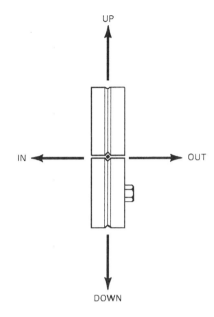

10. What must be pressed, turned, or adjusted to cause the wire drive motor to slowly feed the welding wire inch by inch through the electrode cable? _____

_____

11. The wire and rolls are out of line in the drawing shown below. What must be moved, and in what direction, to correct the alignment? _____

_____

_____

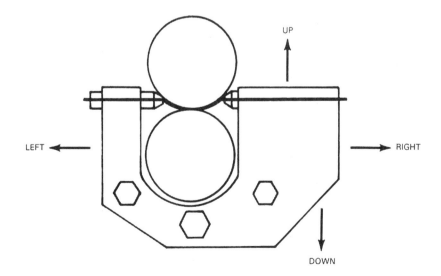

12. The polarity and gas used in GMAW affects the penetration. Name the gas used and polarity used for each of the welds shown below.

    Gas: _____ _____ _____ _____

    Polarity: _____ _____ _____ _____

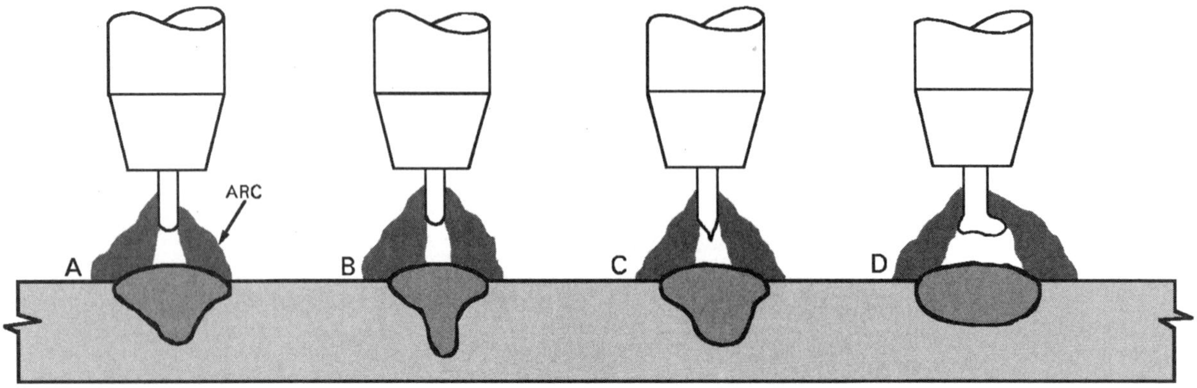

    A   B   C   D

13. Name two gases or gas mixtures used to weld stainless steel. See Figure 15-30 in the text. _____
    _____
    _____

14. Circle the characteristics that apply to argon gas used in GMAW.
    A. Ionizes easily
    B. Quiets the arc
    C. Produces deep penetration
    D. Requires high voltage
    E. Causes a great deal of spatter
    F. Low thermal heat
    G. Less gas is needed
    H. Transfers more heat through the arc

15. Write the letters of those characteristics above which apply to Helium (He) gas.

    15. _____
        _____
        _____

16. _____ cfh (L/min) is the gas flow rate suggested in Figure 15-31 in the text for welding 3/8" (9.5mm) thick aluminum in the overhead welding position?

    16. _____

17. Why must the flow rate be increased when welding in the horizontal, vertical, and overhead welding positions with gases like argon and CO_2? _____
    _____
    _____
    _____
    _____

Lesson 15B GMAW Equipment and Supplies 209

18. Name the parts of the nozzle end of the GMAW torch shown below.

    A. _____
    B. _____
    C. _____
    D. _____
    E. _____
    F. _____
    G. _____
    H. _____
    I. _____

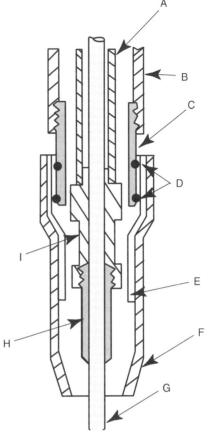

19. Refer to Heading 13.18 in the text. What does the "T" stand for in the electrode number E80T-5?

    19. _____

20. Name the various dimensions in the drawing below.

    A. _____
    B. _____
    C. _____
    D. _____
    E. _____

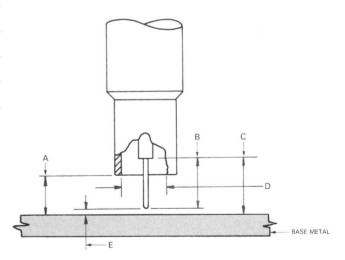

Copyright Goodheart-Willcox Co., Inc.

Job 15B-1

LEARNING OBJECTIVE

★ In this job, you will learn how to turn on a GMAW or FCAW machine. You will also learn how to adjust all the variables required to perform a GMA or FCA weld.

Name _____
Date _____ Class _____
Instructor _____
Grade _____

1. Ask your instructor to assign a GMAW machine to you.

2. Study all the controls, switches, and terminals on your assigned machine. Answer the following questions:

   A. Who is the manufacturer of the machine?   2. A. _____

   B. What is the welding amperage range of the machine?   B. _____

   C. Does the machine have the following? (Circle each correct response.)

1. Remote control receptacle	yes	no
2. Remote contactor receptacle	yes	no
3. Pulses per second control	yes	no
4. Peak amperage control	yes	no
5. Background amperage control	yes	no
6. Remote control switch	yes	no
7. Remote contactor switch	yes	no
8. Ammeter	yes	no
9. Voltmeter	yes	no
10. Power off/on switch	yes	no
11. Wire feed mechanism with a 2 -or 4-drive system	yes	no

   D. Does the wire drive mechanism have the following?

1. An inch switch	yes	no
2. A shielding gas purge switch	yes	no
3. A remote control receptacle	yes	no
4. A wire feed or retract switch	yes	no
5. A wire feed speed control	yes	no

   E. Which wheel is the driving wheel in each pair?   E. _____

   F. Does your station have the following?

1. A regulator	yes	no
2. A flowmeter	yes	no

3. Assume that you are going to GMAW a butt joint in the flat welding position on an aluminum alloy part which is 1/8″ (3.2mm) thick. Assume also that you are using the spray transfer method with a 0.035″ (0.89mm) electrode. The filler wire should be compatible with the aluminum in the base metal. Answer the following questions.

   A. What arc voltage is suggested? (See Figure 15-16)   3. A. _____

   B. What amperage range is suggested?   B. _____

   C. What shielding gases may be used? (See Figure 15-29)   C. _____

   D. What is the suggested argon flow rate?   D. _____
      (See Figure 15-32 in the text.)

4. Carefully examine and study the wire feed mechanism. Refer to Heading 15.3.2 of the text. Be prepared to explain all the switches, controls, and receptacles on the wire drive mechanism. Also be prepared to show how to correct any drive wheel misalignment on the mechanism.

5. Study your gas cylinder, regulator, and flowmeter. Study Headings 5.2.1, 5.2.3, 5.4, 5.4.1, 5.4.6, 5.5, 6.5.6, 6.14.2, 6.14.3 of the text regarding cylinder, regulator, and hose safety. Heading 13.4.3 of the text explains the function and operation of the flowmeter. Be prepared to explain and demonstrate for your instructor how the cylinder is opened and how to properly close down the shielding gas part of the GMAW welding station.

6. Call your instructor when you are prepared to locate and explain the functions of all the controls on your welding machine, gas regulator, flowmeter, and wire feed mechanism. Be prepared to explain and/or demonstrate for your instructor how to load the wire into the wire feeder.

Instructor's initials: _____

Lesson 15C

GMAW AND FCAW PROCEDURES

Name _____ Date _____ Lesson Grade _____
Class _____ Instructor _____

LEARNING OBJECTIVE
★ You will be able to produce welds on butt, lap, inside and outside corners, and T-joints in the flat and horizontal welding positions using GMAW or FCAW.

INSTRUCTIONS
Carefully read Headings 15.4 and 15.6 through 15.9 of the text. Also study Figures 15-41 through 15-50 in the text. Then complete the following problems.

1. Why is it possible and desirable to use a narrower groove when V-groove butt welding with GMAW or FCAW than when welding with SMAW?
   A. The FCAW and GMAW filler wire is thicker.
   B. FCAW and GMAW penetrate more than SMAW.
   C. Narrow groove takes less filler metal.
   D. Narrow groove welds take more welder time to complete.
   E. Narrow groove welds are cheaper to make.

   1. _____

2. In GMAW and FCAW, the welder must control what three variables? _____
   _____
   _____

3. The distance that the welder must control in FCAW or GMAW is called _____.
   A. electrode extension distance
   B. arc length
   C. nozzle to work distance
   D. contact to tube distance
   E. extraneous distance

   3. _____

Copyright Goodheart-Willcox Co., Inc.

4. Identify the drawings below that show the forehand, the backhand, and the perpendicular methods of welding.

   Backhand welding _____

   Forehand welding _____

   Perpendicular welding _____

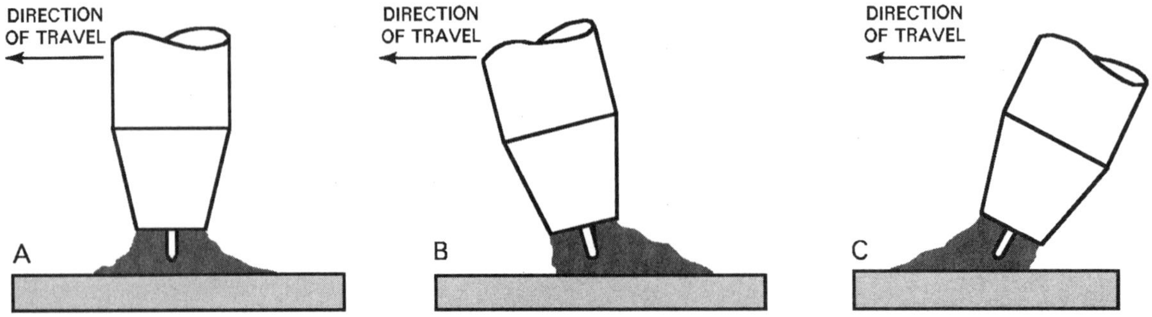

5. Which of the completed welds shown below generally result from the forehand, the backhand, or the perpendicular welding method?

   Backhand welding _____

   Forehand welding _____

   Perpendicular welding _____

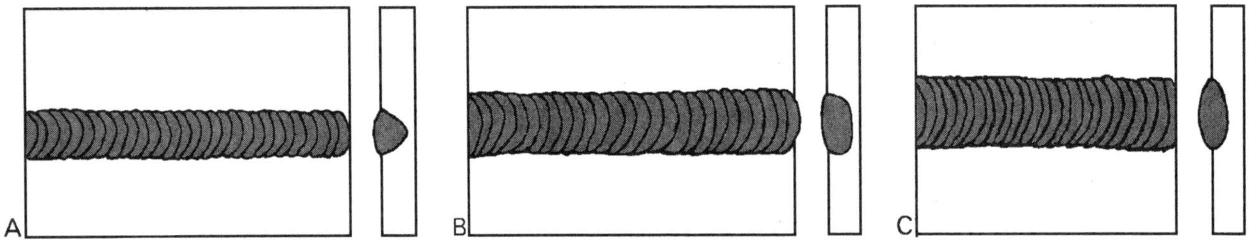

6. In FCAW or GMAW, why is no up-and-down or scratching motion required to strike the arc? _____

   _____
   _____
   _____
   _____

7. *True or False?* A run-off tab may be used to ensure a full width, full penetration weld all the way to the end of the joint.

   7. _____

8. The end of a weld should be shielded while it cools by _____.
   A. squeezing the trigger on the welding gun
   B. shutting off the ventilation
   C. holding the nozzle at the end of the weld during the preflow of the shielding gas
   D. holding the nozzle at the end of the weld during the post flow of the shielding gas
   E. reversing the travel of the gun

   8. _____

9. List the eight steps required to properly shut down an FCA or GMA welding station.

___

10. Place the correct electrode angles on the drawing of the welds shown below. These welds are being made in the flat and horizontal welding positions.

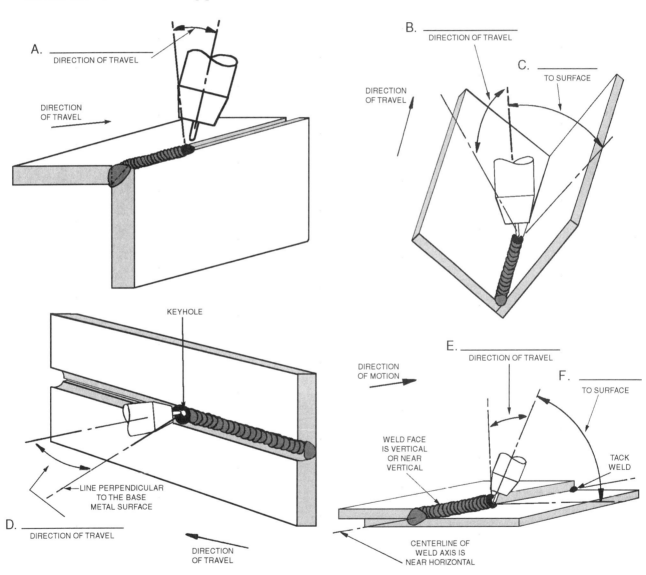

Job 15C-1

LEARNING OBJECTIVE	Name _____
★ In this job, you will demonstrate your ability to turn on the welding machine and set the variables. You will also learn to make a fillet weld on a lap joint in the flat welding position.	Date _____ Class _____ Instructor _____ Grade _____

1. Obtain four pieces of mild steel measuring 1/16" × 1 1/2" × 5" (1.6mm × 38mm × 127mm).

2. A. Use the short circuiting transfer method.

   B. Use an ER70s-X or ER80S-X carbon steel filler wire or electrode or see the latest AWS A.5-18 specification.

   C. The filler wire diameter should be 0.030" or 0.035" (0.76mm or 0.89mm).

   D. Use the shielding gas that is recommended for a cheap, spattery, fast weld with deep penetration. Refer to Figure 15-30 in the text. What shielding gas is suggested for these conditions?
   2. D. _____

3. Before beginning to weld, refer to Figures 15-6, 15-11, and 15-30 in the text and answer the following questions:

   A. What is the minimum spray arc current or transition current for your electrode?
   3. A. _____

   B. Name five shielding gases suggested in the text for use with GMA or FCA welding on mild steel.

   _____
   _____
   _____
   _____
   _____

   C. The arc voltage should be _____ volts.   C. _____

   D. The suggested amperage range should be _____ or _____ amperes.   D. _____

   E. What type current is to be used?   E. _____

See the GMAW gun manufacturer's directions for the correct size of nozzle to use. The general purpose nozzle size for your gun will usually work well for most applications.

4. Set the voltage, wire feed speed (amperage), and dc polarity on the arc welding machine and wire drive mechanism. Open the shielding gas cylinder(s) and set the regulator(s) to the recommended pressure (usually 50 psig). The shielding gas flow rates should be about 30 cfh (ft³./hr.) or 14.16 L/min. After making some test welds, adjust the flow rate so that a rapid crackling sound is heard. When the flow rate is correct, no porosity or discoloring should show in the bead.

5. Make three tack welds in each joint in the weldment shown below.

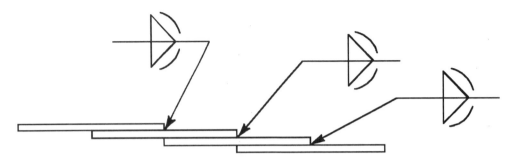

6. Place the weldment into a weld positioning fixture or prop it against firebricks so that all welds can be made in the flat welding position.

7. Complete all welds as shown in #5.

8. Inspect each weld and change whatever needs to be changed so that the next weld is of a higher quality. Increase the gas flow rate if the bead has visible porosity or discoloration.

INSPECTION
Each weld should be convex in shape. The bead should have an even width with evenly spaced ripples. No porosity or bead discoloration should be visible.

Instructor's initials: _____

Job 15C-2

LEARNING OBJECTIVE	
✶ In this job, you will learn how to weld a fillet weld on an inside corner and a T-joint in the flat welding position.	

Name _____
Date _____ Class _____
Instructor _____
Grade _____

1. Obtain the following pieces of mild steel:
   1 piece – 1/8″ × 4″ × 5″ (3.2mm × 38mm × 127mm).
   1 piece – 1/8″ × 1 1/2″ × 4 3/4″ (3.2mm × 38mm × 120.6mm).
   2 pieces – 1/8″ × 1 1/2″ × 4″ (3.2mm × 38mm × 101.6mm).

2. A. Use the spray arc transfer method.

   B. Use an ER70S-X or ER80S-X carbon steel filler wire or electrode or see the latest AWS A.5-18 specification.

   C. The filler wire diameter should be 0.030″ or 0.035″ (0.76mm or 0.89mm).

   D. Use a gas mixture of argon and 2%–5% oxygen for these welds.

3. Before beginning to weld, refer to Figures 15-6 and 15-12 in the text and answer the following questions:

   A. What is the minimum spray arc current or transition current for your electrode? 3. A. _____

   B. The arc voltage should be _____ volts. B. _____

   C. The suggested amperage range should be _____ or _____ amperes. C. _____

   D. What type current is to be used? D. _____

   See the GMAW gun manufacturer's directions for the correct size of the nozzle to use. The general purpose nozzle size for your gun will usually work well for most applications.

4. Set the voltage, wire feed speed (amperage), and dc polarity on the arc welding machine and wire drive mechanism. Open the shielding gas cylinder(s) and set the regulators to the recommended pressure (usually 50 psig). The shielding gas flow rates should be about 30 cfh (ft³/hr.) or 14.16 L/min. After making some test welds, adjust the flow rate so that a rapid crackling sound is heard. When the flow rate is correct, no porosity or discoloring should show in the bead.

5. Tack weld each joint in three places on the weldment below:

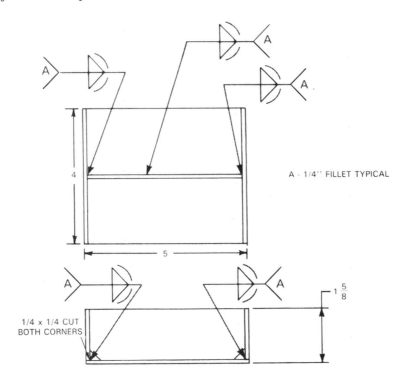

6. Place the weldment into a weld positioning fixture or prop it against firebricks so that all welds can be made in the flat welding position.

7. Complete all welds shown in #5.

8. Inspect each weld and change whatever needs to be changed so that the next weld is of a high quality. Increase the gas flow if the bead has visible porosity or discoloration.

INSPECTION
Each weld should be convex in shape. The bead should have an even width with evenly spaced ripples. No porosity or bead discoloration should be visible.

Instructor's initials: _____

Job 15C-3

LEARNING OBJECTIVE	Name _____
✶ In this job, you will learn to weld a square groove butt joint in the flat welding position.	Date _____ Class _____ Instructor _____ Grade _____

1. Obtain four pieces of mild steel which measure 1/4″ × 1 1/2″ × 5″ (6.4mm × 38mm × 127mm).

2. A. Use the spray arc transfer method.

   B. Use an ER70S-X or ER80S-X carbon steel filler wire, or see the latest AWS A.5-18 specification.

   C. Use a gas mixture of argon and 2% – 5% oxygen for these welds.

   D. The filler wire diameter should be 0.045″ (1.14mm).

3. Before beginning to weld, refer to Figures 15-6 and 15-12 in the text and answer the following questions:

   A. What is the minimum spray arc current or transition current for your electrode?   3. A. _____

   B. The arc voltage should be _____ volts.   B. _____

   C. The suggested amperage range should be _____ to _____ amperes.   C. _____

   D. What type current is to be used?   D. _____

4. Set the voltage, wire feed speed (amperage), and dc polarity on the arc welding machine or wire drive mechanism. Open the shielding gas cylinder(s) and set the regulators to the recommended pressure (usually 50 psig). The shielding gas flow rates should be about 30 cfh (cu.ft./hr.) or 14.16 L/min. After making some test welds, adjust the flow rate so that a rapid crackling sound is heard. When the flow rate is correct, no porosity or discoloring should show in the bead.

5. Tack weld each joint in three places to form the weldment shown below.

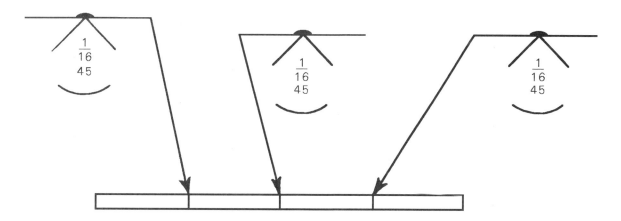

6. Place the weldment into a weld positioning fixture or prop it against firebricks so that all welds can be made in the flat welding position.

7. Complete all welds as shown in #5. **Note:** Watch for and maintain a keyhole at the root of the joint. This will ensure 100% penetration.

8. Inspect each weld and change whatever needs to be changed so that the next weld is of a higher quality. Increase the gas flow rate if the bead has visible porosity or discoloration.

INSPECTION

Each weld should be convex in shape. The bead should have an even width with evenly spaced ripples. No porosity or bead discoloration should be visible. Complete penetration should be visible throughout the entire length of each weld.

Instructor's initials: _____

Job 15C-4

LEARNING OBJECTIVE
* In this job, you will learn to produce a fillet weld on a lap joint in the horizontal welding position.

Name _____
Date _____ Class _____
Instructor _____
Grade _____

1. Obtain four pieces of mild steel that measure 1/8″ × 1 1/2″ × 5″ (3.2mm × 38mm × 127mm).

2. A. Use the spray arc transfer method.
   B. Use an ER70S-X or ER80S-X carbon steel filler wire, or see the latest AWS A.5-18 specification.
   C. The filler wire diameter should be 0.035″ (0.89mm).
   D. Use a gas mixture of argon and 2%–5% oxygen for these welds.

3. Before beginning to weld, refer to Figures 15-6 and 15-12 in the text and answer the following questions:

   A. What is the minimum spray arc current or transition current for your electrode? 3. A. _____

   B. The arc voltage should be _____ volts. B. _____

   C. The suggested amperage range should be _____ to _____ amperes. C. _____

   D. What type current is to be used? D. _____

4. Set the voltage, wire feed speed (amperage), and dc polarity on the arc welding machine or wire drive mechanism. Open the shielding gas cylinder(s) and set the regulators to the recommended pressure (usually 50 psig). The shielding gas flow rates should be about 30 cfh (ft³/hr.) or 14.16 L/min. After making some test welds, adjust the flow rate so that a rapid crackling sound is heard. When the flow rate is correct, no porosity or discoloring should show in the bead.

5. Tack weld each joint in three places to form the weldment shown below. Tack welds may be made in the flat welding position.

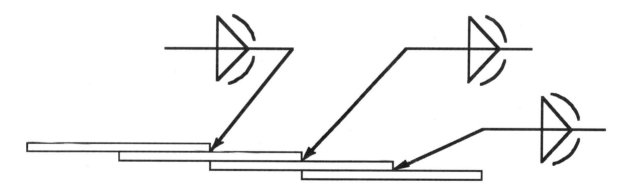

6. Place the weldment into a weld positioning fixture or prop it against firebricks. Make all welds in the horizontal welding position.

7. Complete all welds as shown in #5.

8. Inspect each weld and change whatever needs to be changed so that the next weld is of a higher quality. Increase the gas flow rate if the bead has visible porosity or discoloration.

INSPECTION
Each weld should be convex in shape. The bead should have an even width with evenly spaced ripples. No porosity or bead discoloration should be visible.

Instructor's initials: _____

Job 15C-5

LEARNING OBJECTIVE
* In this job, you will learn to make a fillet weld on a T-joint in the horizontal welding position.

Name_____
Date_____ Class_____
Instructor_____
Grade_____

1. Obtain three pieces of mild steel that measure 1/16" × 1 1/2" × 5" (1.6mm × 38mm × 127mm).

2. A. Use the spray arc transfer method.

   B. Use a gas mixture of argon and 2%–5% oxygen for these welds.

   C. Use an ER70S-X or ER80S-X carbon steel filler wire, or see the latest AWS A.5-18 specification.

   D. The filler wire diameter should be 0.030" or 0.035" (0.76mm or 0.89mm).

3. Before beginning to weld, refer to Figures 15-6 and 15-12 in the text and answer the following questions:

   A. What is the minimum spray arc current or transition current for your electrode?   3. A._____

   B. The arc voltage should be _____ volts.   B._____

   C. The suggested amperage range should be _____ to _____ amperes.   C._____

   D. What type current is to be used?   D._____

4. Set the voltage, wire feed speed (amperage), and dc polarity on the arc welding machine or wire drive mechanism. Open the shielding gas cylinder(s) and set the regulators to the recommended pressure (usually 50 psig). The shielding gas flow rates should be about 30 cfh (ft³/hr.) or 14.16 L/min. After making some test welds, adjust the flow rate so that a rapid crackling sound is heard. When the flow rate is correct, no porosity or discoloring should show in the bead.

5. Tack weld each joint in three places to form the weldment shown below. Tack welds may be made in the flat welding position.

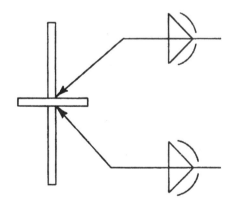

6. Place the weldment into a weld positioning fixture or prop it against firebricks. Make all welds in the horizontal welding position.

7. Complete all welds shown in #5.

8. Inspect each weld and change whatever needs to be changed so that the next weld is of a higher quality. Increase the gas flow rate if the bead has visible porosity or discoloration.

INSPECTION
Each weld should be convex in shape. The bead should have an even width with evenly spaced ripples. No porosity or bead discoloration should be visible.

Instructor's initials: _____

Job 15C-6

LEARNING OBJECTIVE	Name _____
★ In this job, you will demonstrate your ability to produce a square-groove butt joint in the horizontal welding position.	Date _____ Class _____ Instructor _____ Grade _____

1. Four pieces of mild steel will be needed. They should measure 1/8″ × 1 1/2″ × 5″ (3.2mm × 38mm × 127mm).

2. A. Use the spray arc transfer method.

   B. Use an ER70S-X or ER80S-X carbon steel filler wire, or see the latest AWS A.5-18 specification.

   C. The filler wire diameter should be .035″ (0.89mm).

   D. Use a gas mixture of argon and 2% – 5% oxygen for these welds.

3. Before beginning to weld, refer to Figures 15-6 and 15-12 in the text and answer the following questions.

   A. What is the minimum spray arc current or transition current for your electrode?
   3. A. _____

   B. The arc voltage should be _____ volts.
   B. _____

   C. The suggested amperage range should be _____ to _____ amperes.
   C. _____

   D. What type current is to be used?
   D. _____

4. Set the voltage, wire feed speed (amperage), and dc polarity on the arc welding machine or wire drive mechanism. Open the shielding gas cylinder(s) and set the regulators to the recommended pressure (usually 50 psig). The shielding gas flow rates should be about 30 cfh (ft³/hr.) or 14.16 L/min. After making some test welds, adjust the flow rate so that a rapid crackling sound is heard. When the flow rate is correct, no porosity or discoloring should show in the bead.

5. Tack weld each joint in three places to form the weldment shown below. Tack welds may be made in the flat welding position.

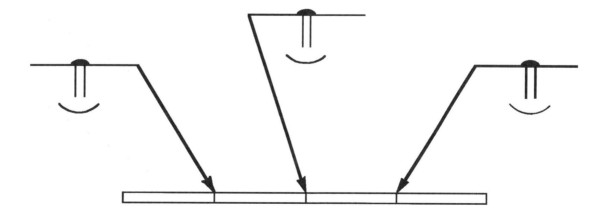

6. Place the weldment into a weld positioning fixture or prop it against firebricks. Make all welds in the horizontal welding position.

7. Complete all welds as shown in #5. **Note:** Watch for and maintain a keyhole at the root of the joint. This will ensure 100% penetration.

Copyright Goodheart-Willcox Co., Inc.

8. Inspect each weld and change whatever needs to be changed so that the next weld is of a higher quality. Increase the gas flow rate if the bead has visible porosity or discoloration.

INSPECTION

Each weld should be convex in shape. The bead should have an even width with evenly spaced ripples. No porosity or bead discoloration should be visible. Complete penetration should be visible throughout the entire length of the weld.

Instructor's initials: _____

Lesson 15D

GMAW AND FCAW PROCEDURES

Name _____ Date _____ Lesson Grade _____
Class _____ Instructor _____

> **LEARNING OBJECTIVE**
> ✶ You will be able to produce welds on butt, lap, inside and outside corners, and T-joints in the vertical and overhead positions using GMAW or FCAW.
>
> **INSTRUCTIONS**
> Carefully read Headings 15.10 through 15.12, 15.13, and 15.14 of the text. Also study Figures 15-51 through 15-58 and Figure 15-63 in the text. Then complete the following problems.

1. When welding in the overhead position, the GMAW metal or FCAW transfer method(s) preferred is(are) _____.
   A. spray transfer
   B. globular transfer
   C. pulsed spray transfer
   D. short circuiting transfer

   1. _____

2. Explain why downhill FCAW is difficult. _____
   _____
   _____
   _____

3. With which metal transfer method is a weaving motion required in order to allow time for the pool or crater to cool?

   3. _____

4. What can be done when the edge melts too quickly while making a fillet weld on a lap joint?
   A. Increase forward speed
   B. Point the electrode more toward the surface
   C. Decrease forward speed
   D. Increase the wire feed rate
   E. Decrease the voltage

   4. _____

Copyright Goodheart-Willcox Co., Inc.

229

5. When doing overhead welds with FCAW or GMAW, why are several narrow beads recommended? _____

6. *True or False?* The angle of the electrode is more vertical when doing an overhead weld.

6. _____

7. A. At what angle to the surface is the electrode inclined in the overhead weld shown below?

   B. At what angle is the electrode tipped in the direction of travel in the drawing below?

7. A. _____

   B. _____

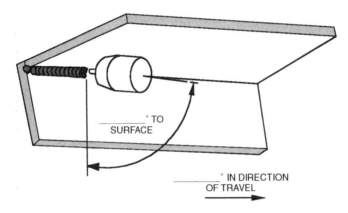

_____° TO SURFACE

_____° IN DIRECTION OF TRAVEL

8. In FCAW or GMAW, explain the difference between semiautomatic and automatic welding. _____

9. When troubleshooting, name four causes of undercut on the base metal. _____

10. How may the condition of excessively wide beads be corrected? _____

Job 15D-1

LEARNING OBJECTIVE	Name _____
★ In this job, you will demonstrate your ability to make a fillet weld on a lap joint in the vertical position.	Date _____ Class _____ Instructor _____ Grade _____

1. Obtain four pieces of mild steel that measure 1/8" × 1 1/2" × 5" (3.2mm × 38mm × 127mm).

2. A. Use the short circuit metal transfer method for these welds.

    B. It is recommended that you use an ER70S-X or ER80S-X carbon steel filler wire or electrode or see the latest AWS A.5-18 specification.

    C. The filler wire diameter for this job should be 0.030" or 0.035" (0.76mm or 0.89mm).

    D. Use a gas mixture of argon and 2% – 5% oxygen for these welds.

3. Before you start to weld, refer to Figures 15-6 and 15-11 in the text and answer the following questions:

    A. What is the minimum transition current for your electrode?   3. A. _____

    B. *Yes or No.* Should you ever set a current above the minimum transition current with GMAW or FCAW with the short circuit metal transfer method?   B. _____

    C. The arc voltage should be _____ or _____ volts.   C. _____

    D. The suggested amperage range should be _____ amperes.   D. _____

    E. What type current is used?   E. _____

See the GMAW gun manufacturer's directions for the correct size of the nozzle to use. The general-purpose nozzle size for your gun will usually work well for most applications.

4. Set the voltage, wire feed speed (amperage), and dc polarity on the arc welding machine and wire drive mechanism. Open the shielding gas cylinder(s) and set the regulators to the recommended pressure (usually 50 psig). The shielding gas flow rates should be about 30 cfh (ft³/hr.) or 14.16 L/min. After making some test welds, adjust the flow rate so that a rapid crackling sound is heard. When the flow rate is correct, no porosity or discoloring should show in the bead.

5. Make three tack welds in each joint in the weldment shown below. Tack weld may be made in the flat position.

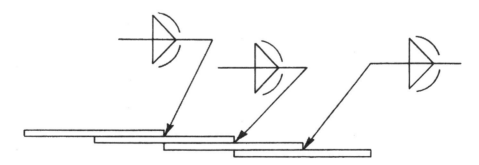

6. Place the weldment into a weld positioning fixture or prop it against firebricks so that all welds are made in the vertical position.
7. Complete all welds as shown in #5.
8. Inspect each weld and change whatever needs to be changed so that the next weld is of a higher quality. Increase the gas flow rate if the bead has visible porosity or discoloration.

INSPECTION
Each weld should be convex in shape. The bead should have an even width with evenly spaced ripples. No porosity or bead discoloration should be visible.

Instructor's initials: _____

Job 15D-2

LEARNING OBJECTIVE
* In this job, you will demonstrate your ability to make a fillet weld on an inside corner and T-joint in the vertical position.

Name _____
Date _____ Class _____
Instructor _____
Grade _____

1. Obtain eight pieces of mild steel which measure 1/16″ × 1 1/2″ × 5″ (1.6mm × 38mm × 127mm).

2. A. Use the pulsed spray transfer method for these welds.

   B. You should use an ER70S-X or ER80S-X carbon steel electrode wire or filler wire, or refer to the latest AWS A.5-18 specification.

   C. The filler wire diameter to use for this job should be 0.030″ or 0.035″ (0.76mm or 0.89mm).

   D. Use a gas mixture of argon and 2% – 5% oxygen for these welds.

3. Before you start to weld, refer to Figures 15-6 and 15-11 in the text and answer the following questions:

   A. What is your electrode's minimum transition current?   3. A. _____

   B. The peak arc voltage should be _____ volts.   B. _____

   C. The suggested peak amperage range should be _____ or _____   C. _____
      amperes for the wire diameter that you are using.

Note: The background voltage must be set 2-3 volts below the lowest suggested peak arc or welding voltage. This lower voltage should be just enough to maintain the arc during the low current or background current part of the welding cycle. See Figure 14-15 in the text for a pulsed arc current diagram. The peak current should be enough to create a good weld buildup and adequate penetration.

   D. What type current is to be used?   D. _____

See the gun manufacturer's recommendation for the correct size nozzle to use. The general-purpose nozzle size for your gun will usually work for most applications.

4. Set the voltage, wire feed speed (amperage), and dc polarity on the arc welding machine and wire drive mechanism. Open the shielding gas cylinder(s) and set the regulators to the recommended pressure (usually 50 psig). The shielding gas flow rates should be about 30 cfh (ft³/hr.) or 14.16 L/min. After making some test welds, adjust the flow rate so that a rapid crackling sound is heard. When the flow rate is correct, no porosity or discoloring should show in the bead.

5. Make three tack welds in each joint in the weldment shown below. Tack welds may be made in the flat position.

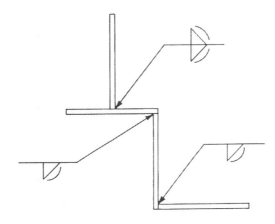

6. Place the weldment into a weld positioning fixture or prop it against firebricks so that all welds are made in the vertical position.

7. Complete all welds shown in #5. Repeat steps #5 and #6 with the remaining pieces of base metal.

8. Inspect each weld and change whatever needs to be changed so that the next weld is of a higher quality. Increase the gas flow rate if the bead has visible porosity or discoloration.

INSPECTION
Each weld should be convex in shape. The bead should have an even width with evenly spaced ripples. No porosity or bead discoloration should be visible.

Instructor's initials: _____

Job 15D-3

LEARNING OBJECTIVE	
★ In this job, you will learn to weld a bevel groove butt joint in the vertical position.	Name _____ Date _____ Class _____ Instructor _____ Grade _____

1. Obtain five pieces of mild steel that measure 1/4″ × 1 1/2″ × 5″ (6.4mm × 38mm × 127mm).

2. A. Use the short circuiting transfer method.

   B. Use an ER70S-X or ER80S-X carbon steel electrode wire or refer to the latest AWS A.5-18 specification.

   C. The filler wire diameter to use for this job should be 0.035″ (0.89mm)

   D. Use a gas mixture of argon and 2% – 5% oxygen for these welds.

3. Before you start to weld, refer to Figures 15-6 and 15-11 in the text and answer the following questions:

   A. What is your electrode's minimum transition current?   3. A. _____

   B. The peak arc voltage should be _____ volts.   B. _____

   C. The suggested peak amperage range should be _____ or _____ amperes for the wire diameter that you are using.   C. _____

Note: The background voltage must be set 2-3 volts below the lowest suggested peak arc or welding voltage. This lower voltage should be just enough to maintain the arc during the low current or background current part of the welding cycle. See Figure 14-15 in the text for a pulsed arc current diagram. The peak current should be enough to create a good weld buildup and adequate penetration.

   D. What type current is to be used?   D. _____

See the gun manufacturer's recommendation for the correct size nozzle to use. The general purpose nozzle size for your gun will usually work for most applications.

4. Set the voltage, wire feed speed (amperage), and dc polarity on the arc welding machine and wire drive mechanism. Open the shielding gas cylinder(s) and set the regulators to the recommended pressure (usually 50 psig). The shielding gas flow rates should be about 30 cfh (ft³/hr.) or 14.16 L/min. After making some test welds, adjust the flow rate so that a rapid crackling sound is heard. When the flow rate is correct, no porosity or discoloring should show in the bead.

5. Make three tack welds in each joint in the weldment shown below. Tack welds may be made in the flat position.

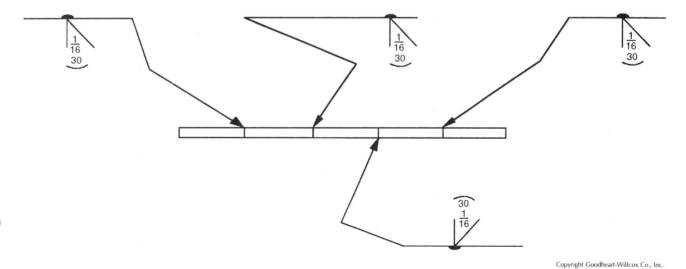

6. Place the weldment into a weld positioning fixture or prop it against firebricks so that all welds are made in the vertical position.

7. Complete all welds shown in #5.

Note: Watch for and maintain a keyhole at the root of the joint. This will ensure 100% penetration.

8. Inspect each weld and change whatever needs to be changed so that the next weld is of a higher quality. Increase the gas flow rate if the bead has visible porosity or discoloration.

INSPECTION

Each weld should be convex in shape. The bead should have an even width with evenly spaced ripples. No porosity or bead discoloration should be visible.

Instructor's initials: _____

Job 15D-4

LEARNING OBJECTIVE
* In this job, you will learn to produce fillet welds on a lap joint in the overhead welding position.

Name _____
Date _____ Class _____
Instructor _____
Grade _____

Caution: Since this weld is made above your head, you must wear a leather cape or jacket on your shoulders, leather spats, and cap on your head to cover your hair. Also observe all other arc welding safety precautions.

1. Obtain four pieces of mild steel that measure 1/16″ × 1 1/2″ × 5″ (1.6mm × 38mm × 127mm).

2. A. Use the pulsed spray transfer method.

    B. Use an ER70S-X or ER80S-X carbon steel electrode wire or refer to the latest AWS A.5-18 specification.

    C. The filler wire diameter to use for this job should be 0.030″ (0.76mm).

    D. Use a gas mixture of argon and 2% – 5% oxygen for these welds.

3. Before you start to weld, refer to Figures 15-6 and 15-11 in the text and answer the following questions:

    A. What is your electrode's minimum transition current? 3. A. _____

    B. The peak arc voltage should be _____ volts. B. _____

    C. The suggested peak amperage range should be _____ or _____ amperes for the wire diameter that you are using. C. _____

Note: The background voltage must be set 2-3 volts below the lowest suggested peak arc or welding voltage. This lower voltage should be just enough to maintain the arc during the low current or background current part of the welding cycle. See Figure 14-15 in the text for a pulsed arc current diagram. The peak current should be enough to create a good weld buildup and adequate penetration.

   D. What type current is to be used? D. _____

See the gun manufacturer's recommendation for the correct size nozzle to use. The general purpose nozzle size for your gun will usually work for most applications.

4. Set the voltage, wire feed speed (amperage), and dc polarity on the arc welding machine and wire drive mechanism. Open the shielding gas cylinder(s) and set the regulators to the recommended pressure (usually 50 psig). The shielding gas flow rates should be about 30 cfh (ft³/hr.) or 14.16 L/min. After making some test welds, adjust the flow rate so that a rapid crackling sound is heard. When the flow rate is correct, no porosity or discoloring should show in the bead.

5. Make three tack welds in each joint in the weldment shown below. Tack welds may be made in the flat position.

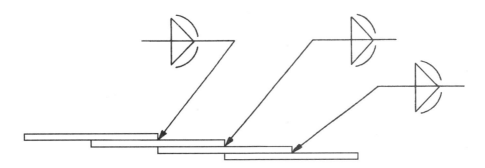

6. Place the weldment into a weld positioning fixture or prop it against firebricks so that all welds are made in the overhead welding position.

7. Complete all welds shown in #5.

Note: Watch for and maintain a keyhole at the root of the joint. This will ensure 100% penetration.

8. Inspect each weld and change whatever needs to be changed so that the next weld is of a higher quality. Increase the gas flow rate if the bead has visible porosity or discoloration.

INSPECTION

Each weld should be convex in shape. The bead should have an even width with evenly spaced ripples. No porosity or bead discoloration should be visible.

Instructor's initials: _____

Job 15D-5

LEARNING OBJECTIVE	Name _____
★ In this job, you will learn to produce a fillet weld on a T-joint in the overhead welding position.	Date _____ Class _____ Instructor _____ Grade _____

Caution: Since this weld is made above your head, you must wear a leather cape or jacket on your shoulders, leather spats, and a cap on your head to cover your hair. Also observe all other arc welding safety precautions.

1. Obtain three pieces of mild steel which measure 1/8" × 1 1/2" × 5" (3.2mm × 38mm × 127mm).

2. A. Use the pulsed spray transfer method for these welds.

   B. You should use an ER70S-X or ER80S-X carbon steel electrode wire or filler wire, or refer to the latest AWS A.5-18 specification.

   C. The filler wire diameter to use for this job should be 0.030" or 0.035" (0.76mm or 0.89mm).

   D. Use a gas mixture of argon and 2% – 5% oxygen for these welds.

3. Before you start to weld, refer to Figures 15-6 and 15-11 in the text and answer the following questions:

   A. What is your electrode's minimum transition current?   3. A. _____

   B. The peak arc voltage should be _____ volts.   B. _____

   C. The suggested peak amperage range should be _____ or _____ amperes for the wire diameter that you are using.   C. _____

Note: The background voltage must be set 2-3 volts below the lowest suggested peak arc or welding voltage. This lower voltage should be just enough to maintain the arc during the low current or background current part of the welding cycle. See Figure 14-15 in the text for a pulsed arc current diagram. The peak current should be enough to create a good weld buildup and adequate penetration.

   D. What type current is to be used?   D. _____

See the gun manufacturer's recommendation for the correct size nozzle to use. The general purpose nozzle size for your gun will usually work for most applications.

4. Set the voltage, wire feed speed (amperage), and dc polarity on the arc welding machine and wire drive mechanism. Open the shielding gas cylinder(s) and set the regulators to the recommended pressure (usually 50 psig). The shielding gas flow rates should be about 30 cfh (ft³/hr.) or 14.16 L/min. After making some test welds, adjust the flow rate so that a rapid crackling sound is heard. When the flow rate is correct, no porosity or discoloring should show in the bead.

5. Make three tack welds in each joint in the weldment shown below. Tack welds may be made in the flat position.

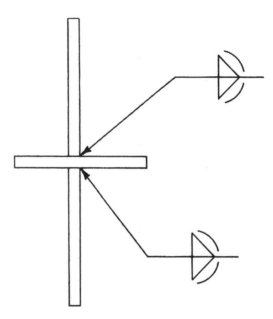

6. Place the weldment into a weld positioning fixture or prop it against firebricks so that all welds are made in the overhead welding position.
7. Complete all welds shown in #5.

Note: Watch for and maintain a keyhole at the root of the joint. This will ensure 100% penetration.

8. Inspect each weld and change whatever needs to be changed so that the next weld is of a higher quality. Increase the gas flow rate if the bead has visible porosity or discoloration.

INSPECTION

Each weld should be convex in shape. The bead should have an even width with evenly spaced ripples. No porosity or bead discoloration should be visible.

Instructor's initials: _____

Job 15D-6

LEARNING OBJECTIVE	Name_____
★ In this job, you will learn to make a square groove butt joint in the overhead welding position.	Date _____ Class _____ Instructor _____ Grade _____

Caution: Since this weld is made above your head, you must wear a leather cape or jacket on your shoulders, leather spats, and a cap on your head to cover your hair. Also observe all other arc welding safety precautions.

1. Obtain four pieces of mild steel that measure 1/16" × 1 1/2" × 5" (1.6mm × 38mm × 127mm).

2. A. Use the pulsed spray transfer method for these welds.

   B. Use an ER70S-X or ER80S-X carbon steel electrode wire or filler wire, or refer to the latest AWS A.5-18 specification.

   C. The filler wire diameter to use for this job should be 0.030" (0.76mm).

   D. Use a gas mixture of argon and 2% – 5% oxygen for these welds.

3. Before you start to weld, refer to Figures 15-6 and 15-11 in the text and answer the following questions:

   A. What is your electrode's minimum transition current?   3. A. _____

   B. The peak arc voltage should be _____ volts.   B. _____

   C. The suggested peak amperage range should be _____ or _____ amperes for the wire diameter that you are using.   C. _____

Note: The background voltage must be set 2-3 volts below the lowest suggested peak arc or welding voltage. This lower voltage should be just enough to maintain the arc during the low current or background current part of the welding cycle. See Figure 14-15 in the text for a pulsed arc current diagram. The peak current should be enough to create a good weld buildup and adequate penetration.

   D. What type current is to be used?   D. _____

See the gun manufacturer's recommendation for the correct size nozzle to use. The general purpose nozzle size for your gun will usually work for most applications.

4. Set the voltage, wire feed speed (amperage), and dc polarity on the arc welding machine and wire drive mechanism. Open the shielding gas cylinder(s) and set the regulators to the recommended pressure (usually 50 psig). The shielding gas flow rates should be about 30 cfh (ft³/hr.) or 14.16 L/min. After making some test welds, adjust the flow rate so that a rapid crackling sound is heard. When the flow rate is correct, no porosity or discoloring should show in the bead.

5. Make three tack welds in each joint in the weldment shown below. Tack welds may be made in the flat position.

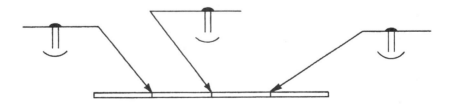

6. Place the weldment into a weld positioning fixture or prop it against firebricks so that all welds are made in the overhead position.

7. Complete all welds shown in #5.

Note: Watch for and maintain a keyhole at the root of the joint. This will ensure 100% penetration.

8. Inspect each weld and change whatever needs to be changed so that the next weld is of a higher quality. Increase the gas flow rate if the bead has visible porosity or discoloration.

INSPECTION

Each weld should be convex in shape. The bead should have an even width with evenly spaced ripples. No porosity or bead discoloration should be visible.

Instructor's initials: _____

Lesson 15E
GMA SPOT WELDING

Name _____ Date _____ Lesson Grade _____
Class _____ Instructor _____

LEARNING OBJECTIVE
★ You will be able to set a GMA spot welding machine for all the variables and perform GMA spot welds.

INSTRUCTIONS
Carefully read Heading 15.13 of the text. Also study Figures 15-36 and 15-60 through 15-62 in the text. Then complete the following problems.

1. GMA spot welding is generally done on metals less than _____ thick.
   A. 1/64″ (0.4mm)
   B. 1/16″ (1.6mm)
   C. 1/32″ (0.8mm)
   D. 1/8″ (3.2mm)
   E. 3/32″ (2.4mm)

   1. _____

2. *True or False?* Spot welds can be made on lap joints, inside corner joints, T-joints, outside corner joints, and on metal surfaces.

   2. _____

3. Identify the variables on the right that are changes to vary the items listed on the left.

   _____ penetration
   _____ spot weld diameter
   _____ welding amperage

   A. arc voltage
   B. electrode extension
   C. wire feed speed
   D. shielding gas
   E. welding time

4. The best type GMA spot welding timer starts timing when the _____.
   A. arc is struck
   B. arc machine is turned on
   C. gun trigger is pulled
   D. shielding gas begins to flow
   E. electrode touches the base metal

   4. _____

243

5. The _____ as the voltage is decreased.
   A. spot weld diameter will decrease
   B. spot weld diameter will increase
   C. penetration will increase
   D. penetration will decrease
   E. arc length will decrease

5. _____

6. A. Which of the spot welding nozzles shown below is used for spot welding outside corners?

6. A. _____

   B. Which of the spot welding nozzles shown below is used for spot welding inside corners?

   B. _____

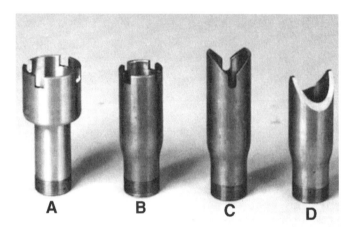

7. List all the variables which must be set on a GMA spot welding machine. _____

_____
_____
_____
_____
_____
_____

8. Name the controls indicated in the drawing at the top of the following page.

   A. _____
   B. _____
   C. _____
   D. _____
   E. _____

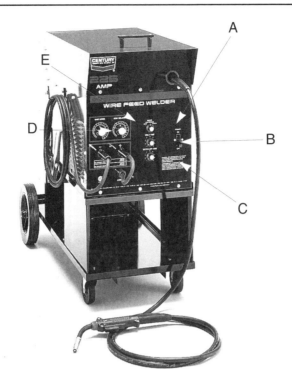

9. *True or False?* Gas metal arc welding and flux cored arc welding may be semi-automatic or fully automatic processes.

9. _____

10. *True or False?* Both GMAW and FCAW guns may be mounted on a motor driven carriage or robot.

10. _____

JOB 15E-1

LEARNING OBJECTIVE	Name _____
★ In this job, you will learn to set the arc welding machine and make GMA spot welds.	Date _____ Class _____ Instructor _____ Grade _____

1. Obtain the following pieces of mild steel:

   6 pieces measuring 0.030″ × 1 1/2″ × 5″ (0.76mm × 38mm × 127mm).

   2 pieces measuring 0.030″ × 3″ × 5″ (0.76mm × 76mm × 127mm).

2. A. Use a GMA welding machine that is capable of making spot welds.

   B. Use an ER70S-X or ER80S-X electrode wire.

   C. The electrode (filler) wire should be 0.030″ (0.76mm) in diameter.

   D. Use CO_2 as the shielding gas.

3. On the spot welding control panel set the following:

   A. The spot, continuous, or pulse control switch to SPOT.

   B. The burnback time control to a little less than halfway in the control range or as suggested by the manufacturer's instruction manual.

   C. The spot time control to a little less than halfway in the time control range or as suggested by the manufacturer's instructions.

   D. The pulse off time control. This control sets the time you require to move to the next spot weld location. When this time elapses (is up), the spot weld cycle will repeat itself if the welding gun trigger is continually held.

   E. The wire speed is set to obtain the amperage required for the electrode wire diameter used.

   F. The welding voltage control. Set for the desired voltage required for the electrode wire diameter used.

   G. Use the short circuit transfer method.

   H. The shielding gas preflow and postflow time should be set. It is often internally set within the welding machine and may not be set by you.

4. Before making any welds, answer the following questions (See Figure 15-11 in the text).

   A. The arc voltage for your wire diameter is _____ volts.   4. A. _____

   B. The amperage range for your wire diameter is _____ amperes.   B. _____

   C. What type current is used?   C. _____

5. Set the dc polarity. Open the shielding gas cylinder and set the regulator to the recommended pressure (usually 50 psig). The shielding gas flow rate should be 30 cfh (ft^3/hr.) or 14.16 L/min.

6. A spot welding nozzle which looks like A or B in Figure 15-36 in the text should be attached to the gun.

7. Use two pieces of 0.030″ × 1 1/2″ × 5″ (0.76mm × 38mm × 127mm) metal. Start the welding machine and place the gun in the center of the base metal about 1″ (25mm) from one end. Hold the gun steady. Pull the trigger. After the welding is completed, release the trigger and continue to hold the gun steady for 10 seconds. This time is required to allow for the gas postflow while the base metal cools.

8. Examine the weld. The spot weld should be round, slightly convex, with a slight melt-through or bumps on the reverse side. The welding engineer or weld drawing will normally give you the desired weld diameter size. Welding machine manufacturer's literature may also suggest the proper weld diameter size.

9. After examining your welding diameter and appearance, change the weld time and/or arc voltage to correct the weld diameter. See Heading 15.14 of the text. Increase the shielding gas flow/rate if the weld appears contaminated.

10. Weld several more spot welds on the same metal as above. Continue to change any variable required until the finished weld looks as you want it to look.

11. Make one spot weld on each of the other two pairs of 0.030″ × 1 1/2″ × 5″ (0.76mm × 38mm × 127mm) metal. These welds should be in the center of the 1 1/2″ (38mm) width and 1″ (25mm) from one end. See below.

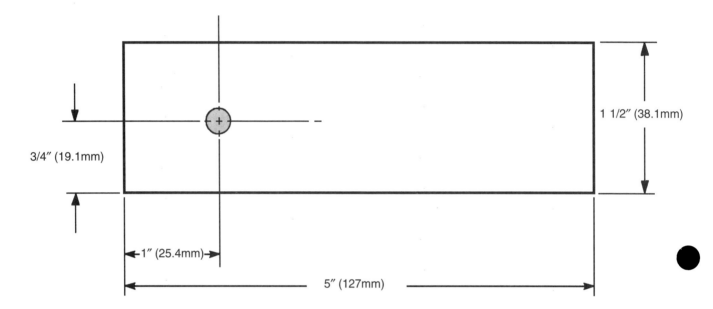

12. Separate the unwelded end and bend them up enough to make a "peel test" in your tensile test.

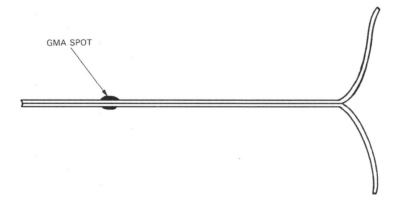

13. Pull the base metal until the weld fails. In a weld with good penetration, the complete weld nugget or weld area will tear out of the base metal. If the weld breaks between the pieces of base metal, the weld penetration is poor.

14. Using the two pieces of 3" (76mm) wide metal, make the welds shown below:

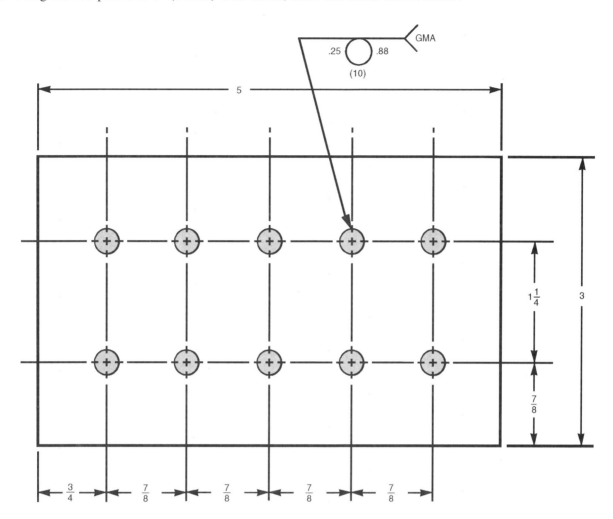

INSPECTION

Your spot welds should be within 1/16" (1.6mm) of the required placement. They should show no signs of contamination. The diameter of the spot weld should be as required by your instructor.

Instructor's initials: _____

Lesson 16A

ARC CUTTING AND GOUGING SAFETY

Name _____ Date _____ Lesson Grade _____
Class _____ Instructor _____

LEARNING OBJECTIVE
✱ You will be able to safely use arc cutting equipment.

INSTRUCTIONS
Carefully read Headings 5.2.1, 5.2.3, 5.4 through 5.4.6, 6.5.1, 6.5.2, 6.5.5, 6.5.6, 6.14.1, 6.14.2, 11.2.2, 11.11, 11.12, 12.4.1, 12.4.2, 12.8, and 13.22 of the text. Then complete the following problems.

1. The oxygen cylinder valve is constructed using a back seating valve to seal the stem form leakage. Therefore, when using the oxygen cylinder, the valve must be _____.
   A. fully closed
   B. half-way open
   C. fully open
   D. whatever is required for the correct pressure
   E. almost open

   1. _____

2. *True or False?* Cylinders must always be kept valve end up. When the cylinder is not in use, the cylinder valve should be closed whether the cylinder is full or empty.

   2. _____

3. Liquid oxygen is stored at _____ °F (_____ °C). If it comes in contact with the eyes or skin, it may cause freeze burns.

   3. _____

4. Name seven organic materials that must be kept away from oxygen in order to prevent fires from occurring.

   _____
   _____
   _____
   _____
   _____
   _____

Copyright Goodheart-Willcox Co., Inc.

5. If clothing has been saturated with oxygen, it becomes highly flammable. It should be removed and not worn again for at least _____ minutes or until no oxygen remains in it.

5. _____

6. *True or False?* Open flames and smoking are not permitted in any area where oxygen is stored, used, or handled.

6. _____

7. Oxygen cylinder valves are closed when they are turned all the way _____.

7. _____

8. Oxygen regulators are closed when the regulator adjusting screw is turned all the way _____.

8. _____

9. *True or False?* Fuel cylinder valve threads are usually left-hand threads. Oxygen cylinder valves usually have right-hand threads. The thread diameters are also different. This is done to prevent connecting the wrong regulator to a cylinder.

9. _____

10. Cylinder valves are protected from damage while they are stored or moved by threading on a properly fitting _____ _____.

10. _____

11. Cylinders in use or in storage should be stored in the vertical position. They should be held upright by a safety chain or metal band. They should be moved using a _____ truck.

11. _____

12. The line voltage to an industrial arc welding machine is generally 220 V or _____. Because of this, only an experienced electrician should work on the electrical power connections used in an arc welding machine.

12. _____

13. The inlet to the ventilation pickup duct should be located so that fumes are removed before they reach the _____ _____.

13. _____

14. Which of the following must be clean and tight to reduce electrical resistance?
    A. Booth curtains
    B. Electrode holder jaws
    C. Work lead at the work
    D. Work lead at the electrode holder
    E. Electrode lead at the machine and holder

14. _____

15. List five primary hazards to avoid when arc welding or cutting. _____
_____
_____
_____
_____

16. *True or False?* You should avoid using arc welding equipment in damp places or with damp welding gloves.

16. _____

17. Select the one statement below which is *not* important when working around arc welding equipment.
    A. Never look at the arc from any distance without wearing an approved filter lens.
    B. Cover all pockets.
    C. Do not carry plastic pens, combs, or matches in your pockets.
    D. Never work without adequate ventilation.
    E. Take every precaution to eliminate H_2O and O_2.

17. _____

18. An arc welding machine must *never* be started under load. Because of this, the electrode holder must be hung on a(n) _____ hanger before the machine is started.

18. _____

19. *True or False?* Safety (chipping) goggles should always be worn when cleaning metals.

19. _____

20. List six pieces of protective equipment or clothing that a welder should wear when arc welding or cutting in the overhead position. _____

_____

_____

_____

_____

_____

_____

21. Which of the following statements is *not* important when the safe handling of oxygen cylinders is involved?
    A. When storing or moving cylinders, the safety cap must be installed.
    B. Cylinders must be securely fastened to a stable object when in use or in storage.
    C. Care must be taken not to accidentally injure cylinders with an arc cutting electrode.
    D. Cylinders should always be used and stored in the upright (vertical) position.
    E. Cylinders must be changed when the pressure in them drops below 250 psig or 1724 kPa.

21. _____

22. *True or False?* If a label on a cylinder is not readable or is missing, you should not assume that a cylinder contains a particular gas. You should return the cylinder to the supplier.

22. _____

23. List four metals which, when welded, create fumes that are irritating or toxic. _____

_____

_____

_____

_____

24. Do not stand in front of the _____ and _____ as the cylinder is opened, even if it is opened slowly.

24. _____
    _____

25. All arc welding and cutting should be done in a booth or in an area protected by curtains to protect others from arc _____ or _____.

25. _____

Lesson 16B

ARC CUTTING AND GOUGING EQUIPMENT AND SUPPLIES

Name _____ Date _____ Lesson Grade _____
Class _____ Instructor _____

LEARNING OBJECTIVE
★ You will be able to identify equipment and supplies used in arc cutting and gouging processes. You will also be able to describe the principles involved in a variety of arc cutting processes.

INSTRUCTIONS
Carefully read Headings 16.1 through 16.6 of the text. Also study Figures 16-1 through 16-27 in the text. Then complete the following problems.

1. Which of the following arc cutting methods does *not* use a shielding gas?
   A. CAC
   B. CAC-A
   C. GMAC
   D. GTAC
   E. SMAC
   F. PAC

   1. _____

2. Which arc cutting process generally has the poorest quality of cut?

   2. _____

3. The polarity used for air carbon arc cutting is _____, and a _____ curve dc welding machine is generally used.

   3. _____

4. To CAC-A cut or gouge to a depth of 1/2" (12.7mm), using a 3/8" (9.5mm) electrode and an N-series carbon torch, what is:

   A. the suggested travel speed? _____

   B. the suggested amperage? _____

5. AOC electrodes are often flux covered to _____.
   A. improve the cutting action with fluxing agents
   B. blow the molten metal out of the kerf
   C. reduce the amperage required
   D. maintain the correct arc length
   E. keep the air from contaminating the electrode

   5. _____

6. The PAC process operates at _____ °F (_____ °C).

   6. _____

Copyright Goodheart-Willcox Co., Inc.

255

7. Name the parts in the arc cutting process shown below.

A. _____

B. _____

C. _____

D. _____

E. _____

F. _____

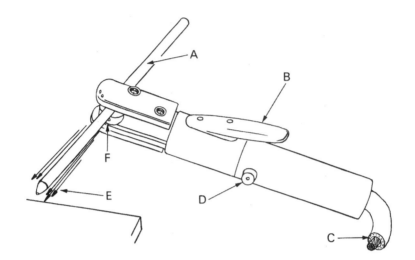

8. The ____ heats the base metal in the water-injected plasma arc process.
   A. heated water
   B. heat of the arc
   C. plasma gas
   D. shielding gas
   E. None of the above.

8. _____

9. Which of the PAC circuits shown below is the:

   A. nontransferred arc process?

   B. transferred arc process?

9. A. _____

B. _____

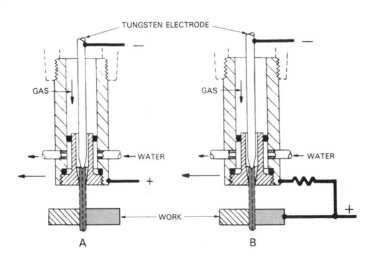

10. Name four gases or gas mixtures which can be used with PAC. _____

_____

_____

_____

_____

Lesson 17

ARC CUTTING AND GOUGING

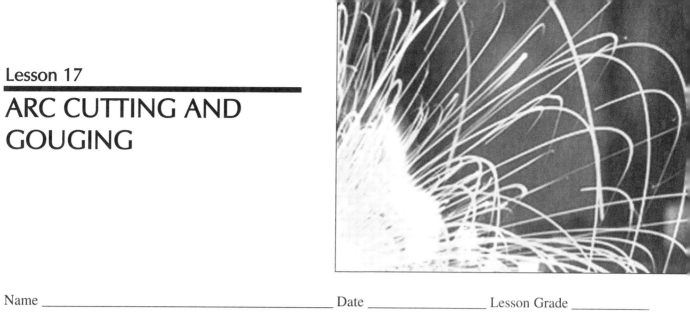

Name _____ Date _____ Lesson Grade _____
Class _____ Instructor _____

> **LEARNING OBJECTIVE**
> ✱ You will be able to cut and gouge using a variety of arc cutting processes. You will also be able to choose and set the variables which produce a quality cut.
>
> **INSTRUCTIONS**
> Carefully read Headings 17.1 through 17.8 of the text. Also study Figures 17-1 through 17-29 in the text. Then complete the following problems.

1. A number _____ shade lens should be used in the arc welding helmet when doing carbon arc cutting.

   1. _____

2. It is best to have the metal in a(n) _____ position when doing CAC.
   A. flat
   B. horizontal
   C. vertical
   D. overhead
   E. It doesn't matter.

   2. _____

3. What two factors must be considered when the welding amperage setting and electrode diameter is chosen? _____

4. When cutting with CAC-A, the air jet holes should be _____ the electrode to get a good cut.

   4. _____

5. Why is a run-on or run-off tab used when arc gouging? _____

6. *True or False?* CAC-A done automatically can hold the depth of a gouge to 0.030″.

   6. _____

Copyright Goodheart-Willcox Co., Inc.

259

7. Suppose you were asked to cut a 1″ (25mm) piece of cast iron using CAC-A. Fill in the following information.

   amps setting on the welding machine   _____

   psig oxygen pressure   _____

   kPa   _____

8. *True or False?* When piercing a hole using SMA, CAC, or CAC-A,   8. _____
   the welder must be especially careful to protect against burns. The
   metal being cut will be driven up out of the hole being cut. Only
   after the hole is cut through, will the molten metal go downward.

9. How is it possible to drag an SMA cutting electrode across the base metal without causing the arc to short out? _____

   _____

   _____

   _____

   _____

10. Unless plasma arc cutting is done under water on a water table, ear   10. _____
    _____ or _____ must be worn.

Job 17-1

LEARNING OBJECTIVE ★ In this job, you will learn to inspect an air carbon arc (CAC-A) welding station.	Name _____ Date _____ Class _____ Instructor _____ Grade _____

The equipment used in a CAC-A station includes the following:

- An ac or dc welding power source.
- An air compressor capable of delivering 80 to 100 psig (551.6 to 689.5 kPa) or a compressed air cylinder.
- A combination electrode lead and air hose.
- Workpiece lead.
- CAC-A electrode holder.
- Ventilation system.
- Booth or screening material.

1. Examine the combination electrode lead and air hose for cuts or worn areas.
2. Check that the compressor filter is clean. If a compressed air cylinder is used, make sure that it is closed.
3. Check that the adjusting knob on the air pressure regulator is turned out and loose, so that the regulator is closed.
4. Check to ensure that the electrode and combination leads are tight on the power source.
5. Hang the CAC-A electrode holder on an insulated hook.
6. Turn on the ventilation system and check that it is working in your area.
7. Ensure that there are no holes in the work booth. If the cutting is done in the open, place protective screening around the work area.
8. Note any discrepancies on this sheet. _____

_____
_____
_____

Instructor's initials: _____

Job 17-2

LEARNING OBJECTIVE
* In this job, you will learn to cut and pierce plain carbon steel using the air carbon arc (CAC-A) process.

Name _____
Date _____ Class _____
Instructor _____
Grade _____

1. Obtain a piece of mild carbon steel that measures 1/2″ × 3″ × 10″ (12.8mm × 76mm × 254mm).

2. Obtain a 1/4″ (6.4mm) air carbon arc electrode.

3. Set up the CAC-A station with the following variables:

   A. Using a DCEP electrode, the amperage should be _____.   3. A. _____
   B. Using an AC electrode, the amperage should be _____.   B. _____
   C. The air pressure fed to the CAC-A holder should be _____. C. _____
   D. The angle of the electrode to the base metal should be _____. D. _____

4. Piercing with the CAC-A process is done in the same manner as is done using SMAC. Reread Headings 17.5 and 17.8 before proceeding.

5. Pierce four holes approximately 1/2″ (12.7mm) in diameter along a line 1/2″ (38mm) from the edge of your plate as shown below.

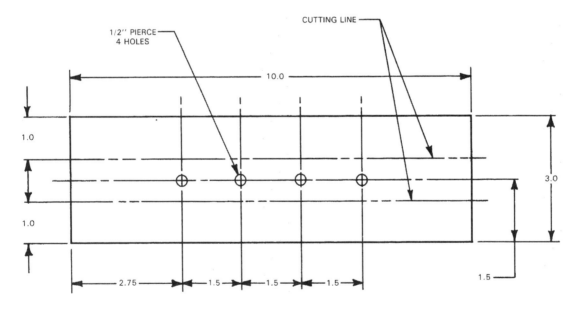

6. Cut the 3″ (76mm) piece of steel used above in two places as shown by the drawing above.

INSPECTION
The edges of the holes and cut surfaces may be rough and ragged. However, they should be as straight as possible. The diameter of the holes and the location of the cuts should be within the tolerance limits set by your instructor.

Instructor's initials: _____

Job 17-3

LEARNING OBJECTIVE
★ In this job, you will learn to remove a weld or the weld reinforcement from a weld bead, using the air carbon arc (CAC-A) process.

Name _____
Date _____ Class _____
Instructor _____
Grade _____

1. Obtain from the scrap container three previously welded pieces of mild steel that are 3/8″ (9.5mm) or thicker. Have the instructor approve your choices.

2. While performing this gouging operation using the CAC-A process, you will use and set the following variables on your cutting station:

   A. Use a 5/16″ (7.9mm) CAC-A electrode.

   B. Using DCEP, the amperage should be _____. 2. B. _____

   C. Using ac, the amperage should be _____.   C. _____

   D. The recommended air pressure is _____ psig or _____ kPa.   D. _____

3. Remove the weld reinforcement from one of your three welds.

4. Check the appearance of the base metal surface to determine if any changes are required to the variables, your technique, or travel speed.

5. Remove the weld reinforcement from the other two welds.

Instructor's initials: _____

Job 17-4

LEARNING OBJECTIVE

★ In this job, you will learn to pierce and cut using the plasma arc cutting (PAC) process.

Name _____
Date _____ Class _____
Instructor _____
Grade _____

1. Obtain pieces of 1/4″ (6.4mm) mild steel, aluminum, and stainless steel that are 3″ (76mm) wide and 10″ (254mm) long.

2. Refer to Figures 17-1, 17-2, and 17-3 to determine the answers to the following:

   To cut the aluminum:

   A. What current setting and polarity is suggested to cut or pierce?   2. A. _____

   B. What orifice diameter is suggested?   B. _____

   C. What is the flow rate suggested for the plasma gas?   C. _____

   D. What is the recommended cutting speed?   D. _____

   To cut the stainless steel:

   A. What current and polarity are suggested to cut or pierce?   A. _____

   B. What orifice diameter is suggested?   B. _____

   C. What is the flow rate suggested for the plasma gas?   C. _____

   D. What is the recommended cutting speed?   D. _____

   To cut plain carbon steel:

   A. What current and polarity are suggested to cut or pierce?   A. _____

   B. What orifice diameter is suggested?   B. _____

   C. What is the flow rate suggested for the plasma gas?   C. _____

   D. What is the recommended cutting speed?   D. _____

3. Mark each piece of metal as shown on the drawing below.

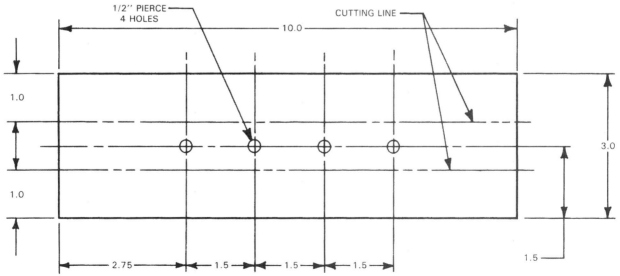

Copyright Goodheart-Willcox Co., Inc.

4. Pierce four holes approximately 1/2″ (12.8mm) in diameter along the line drawn in the middle of each piece.

5. Cut the 3″ (76mm) pieces in two places as shown in the drawing on the previous page.

INSPECTION
The edges of the holes and the edge of the cut metal may be rough. However, they should not be as rough as the edges of an air carbon arc-cut piece. The long cuts should be straight and the holes fairly round. Instructor's initials: _____

Job 17-5

LEARNING OBJECTIVE

* In this job, you will learn to cut a shape on aluminum, stainless steel, and plain carbon steel using the plasma arc cutting (PAC) process.

Name _____
Date _____ Class _____
Instructor _____
Grade _____

1. Obtain a 1/4" (6.4mm) piece of aluminum, stainless steel, and plain carbon steel that is 2" (50.8mm) wide and 6" (152.4mm) long.
2. Using the variables used in Job 17-4, cut the shape shown in the drawing below.
3. Use soapstone to draw the shape shown below on each piece of base metal.

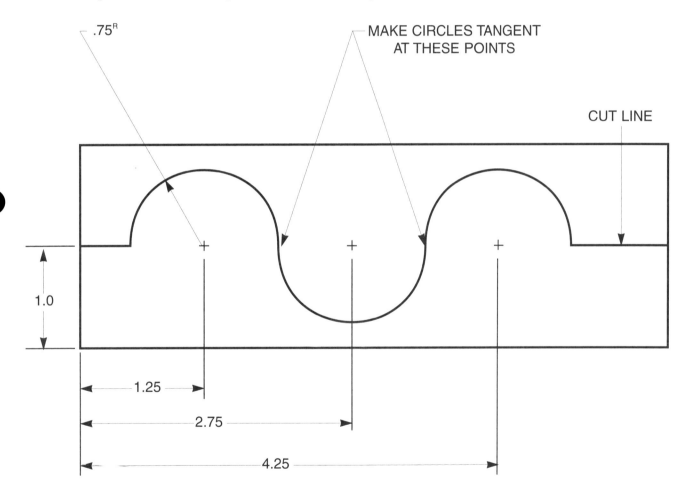

4. Cut along the line drawn on each piece of base metal.

INSPECTION
The edges of the shape you cut should be square and fairly smooth. The shape should be cut as close to the line drawn as possible.

Instructor's initials: _____

Lesson 18A

RESISTANCE WELDING SAFETY

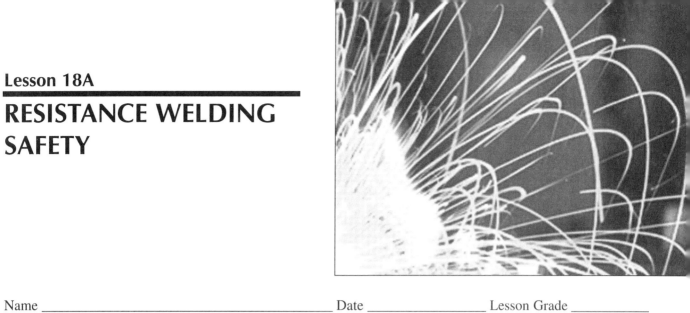

Name _____ Date _____ Lesson Grade _____
Class _____ Instructor _____

LEARNING OBJECTIVE
✷ You will be able to use resistance welding equipment safely.

INSTRUCTIONS
Carefully read Headings 18.14, 19.2.3, 19.2.4, 19.6, and 19.12 in the text. Also study Figures 19-4, 19-13, 19-14, 19-15, 19-30, 19-31 and 19-32 in the text. Then complete the following problems.

1. *True or False?* When molten metal squirts out from a spot weld, it is called an expulsion weld.

    1. _____

2. _____ should be worn to prevent burns and eye injury resulting from flying molten metal and from flash welding.
    A. Safety glasses
    B. Flash goggles or face shield
    C. Gloves
    D. Protective clothing
    E. All of the above.

    2. _____

3. In general, _____ should be worn when handling spot welded metal and sheet metal.

    3. _____

4. List two hazards that may result from an expulsion spot weld. _____
_____
_____

5. The force system of a resistance welding machine can cause great bodily harm. One way to prevent an operator's hand from getting caught between the electrodes is to use _____ _____ switches. (See Text Heading 19.2 and Figure 19-4.)

    5. _____

6. What does the abbreviation OSHA mean? _____
_____
_____

7. If work is to be done on a welding machine secondary, which of the following should be done? (More than one answer may be correct.)
   A. Turn off the resistance welding machine.
   B. Turn off the power at the substation.
   C. Turn off the machine's circuit breaker.
   D. Remove the fuses from the machine.

7. _____

8. If work must be done on the primary, what should be done? _____
_____
_____

9. If the power must be disconnected at the substation, a sign should be placed on the circuit breaker. What information should the sign contain? _____
_____
_____

10. If a hydraulic leak develops, you should _____.
    A. examine the machine to determine where the leak is
    B. tell everyone in the class of the problem
    C. turn off the machine and tell instructor or supervisor
    D. repair the leak by wrapping with black electrical tape

10. _____

Lesson 18B

RESISTANCE WELDING MACHINES

Name _____ Date _____ Lesson Grade _____
Class _____ Instructor _____

> **LEARNING OBJECTIVE**
> ★ You will be able to describe the types of resistance welding machines, their major components, and the types of welds produced.
>
> **INSTRUCTIONS**
> Carefully read Headings 18.1 to 18.3, 18.8 to 18.13, 19.1, 19.1.1, 19.8, 19.9, and 19.10 of the text. Review Heading 4.17 of the text. Also study the Figures referred to under these Headings in the text. Then complete the following problems.

1. List the main parts of a resistance welding machine. _____
   _____
   _____

2. *True or False?* The secondary windings of a resistance welding machine are usually water-cooled, but small transformers can be air-cooled.

   2. _____

3. A transformer capacity is listed as a _____ rating.
   A. VA
   B. KVA
   C. percent heat
   D. duty cycle

   3. _____

4. Label the parts of the transformer shown below.

   A. _____
   B. _____
   C. _____
   D. _____

5. The term _____ is used to describe the amount of time a transformer delivers current in ratio to the time the current is off.
   A. KVA
   B. percent heat
   C. duty cycle
   D. tap setting

5. _____

6. A resistance welding transformer is usually rated at _____ percent duty cycle. The rating is based on a _____ minute time period.

6. _____

7. A transformer that increases the supplied current and decreases the supplied voltage is called a _____ transformer.

7. _____

8. Identify the primary and secondary windings in the transformer drawing below:

   A. _____
   B. _____

9. The selection of a welding machine determines what three variables? _____

_____

_____

10. List the three ways a resistance welding machine can obtain electrical energy. _____

_____

_____

11. A three-phase machine has _____ SCRs and _____ primary transformers. (Fill in the blanks with a number.)

11. _____
_____

12. When making multiple projection welds, _____ (greater, lesser, or equal) forces and _____ (greater, lesser, or equal) current are required, compared to single projection welds.

12. _____
_____

13. The three electrode drive mechanisms used on seam welding machines are _____ drive, _____ drive, and _____ drive.

13. _____
_____
_____

14. *True or False?* A seam welding machine that has the wheel electrodes perpendicular to the front of the machine will produce a transverse seam.

14. _____

15. The dimension marked "A" below is called the _____ _____ and controls the _____ of a part to be welded.

    15. _____
        _____

    The dimension marked "B" is called the _____ _____ and controls the _____ of a part to be welded.

        _____
        _____

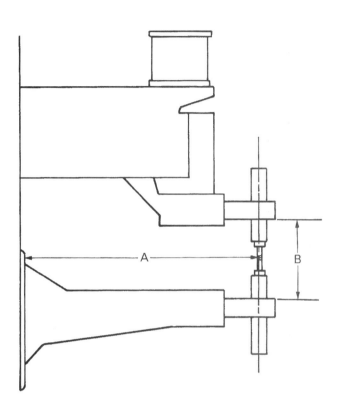

16. Complete the sketch below to indicate how a completed upset weld and a completed flash weld would appear.

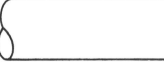

UPSET WELD

FLASH WELD

17. Most spot welding requires the joint to be a _____ joint. When metal foil welding, a _____ joint can be welded.

    17. _____
        _____

18. Where does a high frequency current flow in a conductor? _____
_____
_____

274 Modern Welding Lab Manual

19. What type of machine is shown in #20 below? _____

20. Label the parts of the machine shown below.

　　A. _____
　　B. _____
　　C. _____
　　D. _____
　　E. _____
　　F. _____
　　G. _____
　　H. _____

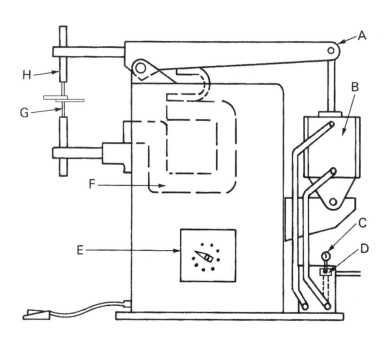

Lesson 18B Resistance Welding Machines 275

Job 18B-1

LEARNING OBJECTIVE

★ In this job, you will learn the major components of the resistance spot welding machines in your shop. Your instructor will select a machine to use for this and the following assignment. This job assignment will be done as an instructor demonstration with student participation. A group of students may work together. *The machine will remain off during this assignment.*

Name _____
Date _____ Class _____
Instructor _____
Grade _____

1. Read Heading 18.14 of the text and listen to the safety instructions given by your instructor. Check and make sure the machine is off. Your instructor may also remove the fuses or turn off the circuit breaker for added safety.

2. Who is the manufacturer of the machine you are studying? _____
_____

3. What is the machine's KVA rating? (See the nameplate.) _____
_____

4. What type of machine is it? 4. _____
   A. Rocker arm
   B. Press
   C. Portable
   D. Seam
   E. Other

5. Your instructor will open the controller cabinet or the side of the machine and point out the main power connections. Never open the controller cabinet or side when the power is on. How many ignition tubes or SCRs are there? 5. _____

6. How does the machine obtain its electrical energy? 6. _____
   A. Single-phase machine
   B. Three-phase machine
   C. Stored energy machine

7. Answer either question A or B. (Whichever applies to the machine you are using.)

   A. What model ignition tubes are used in the machine? 7. A. _____
      What is the current capacity of this model? _____

   B. What is the current capacity of the SCRs used in the machine? B. _____

8. Your instructor will point out the transformer on the machine and show you the main power connection to the transformer. Measure the height, length or depth, and width of the transformer.

Height: _____

Length: _____

Width: _____

Copyright Goodheart-Willcox Co., Inc.

9. Measure the throat depth and horn spacing of the machine.

Throat depth: _____

Horn spacing: _____

10. Can either of the dimensions listed in #9 be changed? If so, how? If not, why not? _____

_____

_____

_____

INSPECTION
Your instructor will point out the safety features on the machine. Your instructor will close the control cabinet or side of the machine when the discussion of the parts within the control cabinet has been completed. You and your instructor should check to make sure the machine is in a good safe working condition before you leave it.

Instructor's initials: _____

Lesson 18C

RESISTANCE WELDING ELECTRICAL COMPONENTS

Name _____ Date _____ Lesson Grade _____
Class _____ Instructor _____

LEARNING OBJECTIVE
★ You will be able to describe the electrical components of a resistance welding machine, including the electrodes and electrode holders. You will also be able to explain how the cooling system of a resistance welding machine operates.

INSTRUCTIONS
Carefully read Headings 18.5 to 18.7 of the text. Also study Figures 18-4 through 18-26 in the text. Then complete the following problems.

1. What theory states that current flows from positive to negative? _____
_____

2. Label the parts on the SCR shown below and indicate with an arrow the direction of the current flow.

   A. _____
   B. _____ (+ or –)
   C. _____
   D. _____ (+ or –)
   E. _____

3. *True or False?* To vaporize the mercury in an ignition tube, a small current must flow from the cathode to the ignitor.

3. _____

4. The amperage capacity of an ignition tube model B is _____. _____ cooling water (gallons/minute) is required.

4. _____
   _____

5. Each SCR will conduct electricity _____ times per second.
   A. 30
   B. 60
   C. 90
   D. 120

5. _____

278   Modern Welding Lab Manual

6. Label the parts of the resistance welding circuit shown.

   A. _____
   B. _____
   C. _____
   D. _____
   E. _____
   F. _____
   G. _____
   H. _____
   I. _____
   J. _____
   K. _____

7. List four requirements of resistance welding electrodes. _____

   _____
   _____
   _____
   _____

8. What group and class of electrode is used for high production spot welding of mild steels and stainless steels?

   _____

9. Label the parts of the electrode shown below. With arrows, show the direction of waterflow through the electrode.

   A. _____
   B. _____
   C. _____

10. List the three ways the electrode or adaptor can be attached to the electrode holder. _____
_____
_____
_____

Job 18C-1

> **LEARNING OBJECTIVE**
> ★ In this job, you will learn how the resistance spot welding machine is cooled. You will also learn how the electrodes are constructed. Your instructor will select a machine on which to perform this assignment. ***The machine will remain off during this assignment.***

Name _____
Date _____ Class _____
Instructor _____
Grade _____

1. Follow your instructor's safety instructions. Do not turn the machine on during this assignment. Your instructor will work with you when performing this job assignment. Students can work in groups.

2. List the manufacturer and KVA rating of the machine you are studying in this assignment. _____
   _____

3. On the nameplate of most resistance welding machines, specific information is given on minimum water pressure and maximum water temperature. Find this information and copy it below. **Note:** If information is not available, write "not available." _____
   _____
   _____
   _____

4. List the parts of a resistance welding machine that are usually cooled. _____
   _____

5. Locate the water inlet on the machine. From this starting point, follow the flow direction of the cooling water through the machine by tracing the hoses. After cooling the machine parts, the water will leave the machine. In the space below, sketch the path the cooling water followed in your machine. Sketch the parts of the machine that were cooled and any valves and/or branches along the path of the water hoses.

WATER INLET

WATER OUTLET

6. Are each of the parts of the machine listed in question #4 cooled in the machine you examined?

   If not, list those parts not water-cooled. _____

   6. _____

7. Turn off the water to the machine. Remove one of the electrodes from the electrode holder. If your machine has an adaptor and cap, take them apart. Have a large cup and a towel available to catch any water that leaks when the electrode is removed. In the space below, sketch the electrode that you removed from your machine. Measure the length, the diameter, and any other important features and write the dimensions on your sketch. If the electrode has a cap and an adaptor, sketch and dimension both.

8. What type of electrode face does the electrode have? _____

9. To determine how close to the end of the electrode the cooling water flows, do the following: (If the hole is deep, insert a piece of 0.035" or 0.045" wire down to the bottom of the hole. Mark the wire even with the top of the electrode. Remove the wire and measure from the mark to the end of the wire. Enter this value in line B below. Subtract the depth of the hole [line B] from the length of the electrode or cap [line A]. This is how close the water comes to the end of the electrode.)

   A. Measure the length of the electrode or the cap.

   B. Measure the depth of the hole where the water flows.

   C. How close to the end of the electrode does the cooling water flow?

   D. Subtract line A from line B to get the answer.

   9. A. _____

   B. _____

   C. _____

   D. _____

Instructor's initials: _____

Lesson 19

SET-UP AND WELDING

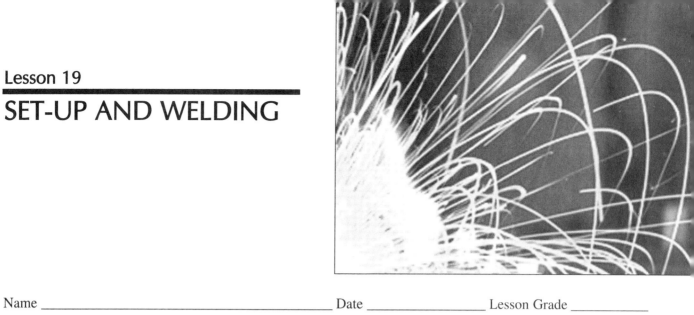

Name _____ Date _____ Lesson Grade _____
Class _____ Instructor _____

> **LEARNING OBJECTIVE**
> * You will be able to determine the proper settings for resistance welding mild steel. You will be able to set up the machine for welding. You will also be able to produce spot welds.
>
> **INSTRUCTIONS**
> Carefully read Headings 18.4, 19.2, 19.2.3, 19.2.4, and 19.2.5 of the text. Also study Figures 18-10 through 18-13, Figures 19-2 through 19-4, and 19-7, and 19-18 in the text. Then complete the following problem.

1. Give a brief definition of the following terms.

    A. Weld time _____
    _____

    B. Hold time _____
    _____

    C. Squeeze time _____
    _____

    D. Off time _____
    _____

2. Give a brief definition of the following terms.

    A. Force gauge _____
    _____

    B. Current analyzer _____
    _____

    C. Controller _____
    _____

3. *True or False?* The tap switch on the transformer is used to make large changes in the secondary current.   3. _____

Copyright Goodheart-Willcox Co., Inc.

4. The _____ is used to make small changes in current.
   A. duty cycle
   B. KVA rating
   C. percent heat control
   D. weld time

4. _____

5. On the weld schedule below, label the axes A, B, and C. Identify the areas marked D, E, and F.

5. A. _____
   B. _____
   C. _____
   D. _____
   E. _____
   F. _____

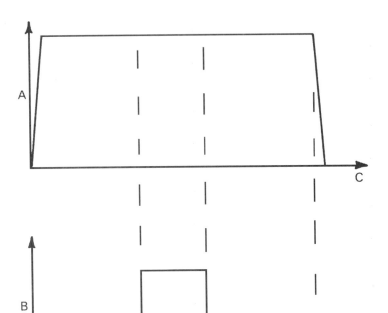

6. Develop a welding schedule using one piece of 0.060" (1.52mm) thick mild steel and one piece of 0.040" (1.02mm) thick mild steel. (See Heading 19.2.3 and Figure 19-7 in the text.)

   A. What contact tip diameter should be used?
   B. What weld time should be used?
   C. What welding current should be used?
   D. What electrode force should be used?
   E. What size should the complete weld be?

6. A. _____
   B. _____
   C. _____
   D. _____
   E. _____

7. Fill in the following graph and draw a weld schedule using answers from #6. Use a 20 cycle squeeze time and a 30 cycle hold time.

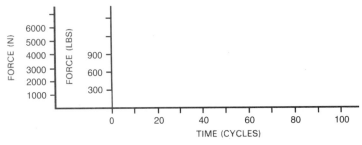

8. Fill in the following graph and draw a weld schedule using answers from #6. Use a 20 cycle squeeze time and a 30 cycle hold time.

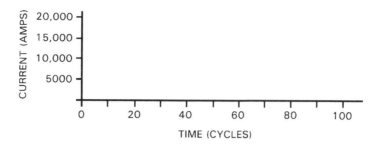

9. How is the desired electrode force set when using an air operated (pneumatic) force system? _____
_____

10. After welding _____ spot welds on mild steel, the electrodes begin to flatten out. It is necessary to reshape the electrode, which is also called _____ the electrode.

10. _____
    _____

Job 19-1

> **LEARNING OBJECTIVE**
> ✶ In this job, you will learn to set all the variables required to make a high quality spot weld. You will also make and test several spot welds. Your instructor will assign a machine to you on which to perform this assignment.

Name _____
Date _____ Class _____
Instructor _____
Grade _____

Who manufactures the machine you are using? _____
_____

Obtain ten pieces of mild steel about 2″ × 8″ (50mm × 203mm). All pieces should be the same thickness, but they do not have to be. On each set of two pieces, 6 to 8 spot welds will be made. If the metal is rusty or has grease, oil, or paint on it, it must be cleaned. Use a degreasing solution and/or a wire brush to remove all of these from the metal.

1. How thick is the metal you will be welding?

    1. _____

2. From Figure 19-7 in the text, determine the following:

    A. What tip diameter should be used?

    B. What weld time should be used?

    C. What weld current should be used?

    D. What weld electrode force should be used?

    E. What weld size should be produced?

    2. A. _____
    B. _____
    C. _____
    D. _____
    E. _____

Show your calculations below.

3. Select an electrode or electrode cap close to the size calculated in #2 above.

    A. What is the actual diameter?

    B. What face design does the electrode have?

    3. A. _____
    B. _____

4. Install or attach the electrode or electrode cap to the welding machine. (See Heading 19.2.4 of the text.)

5. Turn on the welding machine. Turn on the water. Find the switch marked "WELD – NO WELD." Place it in the "NO WELD" position. This allows all the functions of the machine to operate but no electrical current will flow through the electrode.

6. Set a 30 to 60 cycle squeeze time, 0 cycles weld time, and a 30 cycle hold time on the controller. If a force gauge is not available, use the following technique to set the pressure. Place two pieces of the material to be welded between the electrodes. Press the foot control or palm switch to make the electrodes close on the metal. If no indentation is made on the metal, increase the electrode force by increasing the pneumatic (air) or hydraulic pressure, or by increasing the spring tension. Continue increasing the electrode force and cycling through additional weld sequences until the electrodes indent the metal slightly. Next, reduce the electrode force slowly until no indentation occurs. (If this force still seems to be high, reduce the force slightly.) If a force gauge is available, use it to set the correct electrode force. Adjust the valves on a pneumatic (air) or hydraulic machine to obtain the correct force and electrodes. Adjust the spring tension on a manual force machine to obtain the correct force.

7. The air pressure or hydraulic pressure is set at _____ psig. (No answer required if a spring force machine is used.)

7. _____

8. When spot welding, there is a minimum distance that must be maintained between spot welds. The following charts lists the distance based on metal thickness.

THICKNESS OF THE THINNEST PIECE IN INCHES	.021	.031	.040	.050	.062	.078	.093	.109	.125
MINIMUM WELD SPACING IN INCHES	3/8	1/2	3/4	7/8	1	1 1/4	1 1/2	1 5/8	1 3/4

9. What spacing should be used for the thickness of metal you are using? **Note:** Use this weld spacing for the following welding tasks.

9. _____

10. On the controller, set the weld time determined in question #2. Switch the Weld–No Weld switch to Weld. Set the weld current. Use the methods described in #11 or #12. The machine is now ready to weld.

11. To determine the current setting by the trial and error method, proceed as follows:

    A. Set the tap switch to a value less than half the highest value. If there are eight tap settings, select number 3. If there are six tap settings, select number 2. If there are five or less settings, select number 1.

    B. Set the percent heat control at 70%.

    C. Place two pieces of metal to be welded between the electrodes.

    D. Press the foot control or palm switch to begin the welding sequence.

    E. If an expulsion weld occurs, reduce the tap setting by two. If the two pieces are not welded at all or if they can be torn apart by hand, increase the tap setting by one.

    F. Provided neither of these two conditions exist, the weld must be torn apart to examine the weld size. The technique for peel testing a weld is described at the end of this job assignment. Follow the steps described there and measure the weld nugget size. If the weld is less than 75% of the electrode diameter, increase the current by increasing the tap setting on the percent heat control.

G. Continue welding additional welds following steps #4, #5, and #6 until the weld size is 75% to 95% of the electrode diameter. When the weld size is 75% to 95% of the electrode diameter, the current is correct. Try to obtain a weld size of 90% of the electrode diameter. Each spot weld will vary slightly in size, so weld three or more samples to make sure the size remains nearly the same.

Note: If the percent heat control is at 90% or above, and you need just a little more current, do the following: Increase the tap setting by one number and reduce the percent heat control to 50%. If the percent heat control is at 40% or less, and you need slightly less current, do the following: Decrease the tap setting by one and increase the percent heat control to 85%.

12. To set the current, using a current analyzer, proceed as follows:

    A. Follow step one from the trial and error method.

    B. Follow step two from the trial and error method.

    C. Place the pick-up coil around the lower electrode and turn on the current analyzer.

    D. Follow step #3 from the trial and error method.

    E. Follow step #4 from the trial and error method.

    F. The welding current measured in amps or kiloamps (1000 amps) will be displayed on the current analyzer.

    G. Adjust the tap setting and percent heat control to obtain the desired current. If the value on the current analyzer is too low, increase the tap setting or percent heat control. If the value is too high, decrease the tap setting or percent heat control. See the note under trial and error method.

    H. Repeat steps #4 through #7 until the current analyzer reads the correct value.

    I. Once the correct current is set, make a spot weld and peel test it. Check the diameter of the weld to make sure the current set will produce a weld of proper size. If the weld size is not correct, make adjustments in the current to produce a weld of correct size. Follow steps D, E, F, and G from the trial and error method.

13. What tap setting is used? 13. _____

    The heat control is set at _____ %. _____

14. Your machine should now be properly set to weld.

15. Take two pieces of 2" × 8" (50mm × 203mm) mild steel and overlap them about 7 1/2" (191mm). Start 1 1/2" (38mm) from one end. Make as many welds as possible down the center of the two pieces, maintaining the minimum weld spacing from #5.

16. Weld three sets of material together. Tear them apart using the techniques described under "inspection" on page 291.

17. Measure the sizes of two welds from each piece and list the sizes here. Notice the variation even when no change is made in the machine settings.

    1. _____
    2. _____
    3. _____
    4. _____
    5. _____
    6. _____

18. In the next few steps, the effects of weld spacing will be studied. Previously, welds were made using the minimum spacing shown on the weld spacing chart. Welds made for this task will be made at one-half or less the spacing shown in the chart.

19. What is one-half of the minimum weld spacing shown in the table for the thickness of material that you are using?   19. _____

20. Use the same machine settings as before. Weld two pieces of material together with 8 to 12 spot welds. The first weld should be 1 1/2″ (38mm) from one end. The weld spacing should be equal to or less than the spacing from #19. Number the welds. The first weld is #1, the second is #2, etc. Tear the two pieces of metal apart.

21. Measure the diameter of each spot weld and record the size.

    1. _____
    2. _____
    3. _____
    4. _____
    5. _____
    6. _____
    7. _____
    8. _____
    9. _____
    10. _____
    11. _____
    12. _____

22. Does the weld size increase, decrease, or remain the same as each weld is made? **Note:** To prevent this from happening when spot welding, always maintain the minimum weld spacing. The same thing occurs when the welding electrodes begin to wear. To prevent this, the electrodes must be dressed regularly.   22. _____

INSPECTION

Technique for Peel-Testing Spot Welds

Read Heading 30-18 of the text. A peel test separates the two pieces of metal spot welded together. The spot weld usually tears out one piece and can be measured on the other piece. (This does not always occur on thicker pieces of metal.)

Follow these steps to peel test a spot weld:

1. Bend back the 1 1/2" (38mm) portion of the metal that is not welded. One piece of metal will go one way 90°. The other piece will be bent 90° the other way. These can be pulled apart to examine the spot weld. See the sketch below and Figure 30-43 in the text.

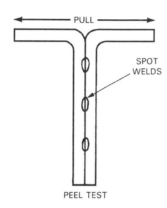

2. Place each bent piece in one jaw of a tensile test machine. A tensile test machine is shown in Figure 30-26 in the text. Tighten each jaw onto the metal.

3. Apply a force to pull the pieces apart.

4. When the spot welds tear out of one piece of metal, stop pulling and remove each piece of metal.

5. Measure the diameter of the spot welds. Use a set of calipers. If no calipers are available, use a scale and eyeball the diameter. Sizes ____ ____ ____.

5. _____
   _____
   _____

Instructor's initials: _____

Job 19-2

LEARNING OBJECTIVE

★ In this job, you will learn to set all the variables required to make a high-quality spot weld. You will also make and test several spot welds. This assigned job is the same as Job 19-1 except that you should use a different thickness of material or a different machine. All instructions are the same, so only the questions are repeated in this assigned job.

Name _____
Date _____ Class _____
Instructor _____
Grade _____

1. Who is the manufacturer of the machine you are using? _____
   _____

2. How thick is the metal you will be welding? _____
   _____

3. From Figure 19-7 in the text, determine the following:

   A. What tip diameter should be used?   3. A. _____
   B. What weld time should be used?   B. _____
   C. What weld current should be used?   C. _____
   D. What electrode force should be used?   D. _____
   E. What weld size should be produced?   E. _____

   Show your calculations below.

4. What is the actual diameter of the electrode selected?   4. _____

5. The air pressure or hydraulic pressure is set at _____ psig. (No   5. _____
   answer required if a spring force machine is used.)

6. What spacing should be used for the thickness of metal you are   6. _____
   using?

7. What tap setting is used?   7. _____

   The heat control is set at _____ %.

8. After spot welding using the previously determined values, measure the weld size and list the sizes here. Notice the variation even when no change is made in the machine settings.

   1. _____
   2. _____
   3. _____
   4. _____
   5. _____
   6. _____

9. What is one-half of the spacing shown in the table for the thickness of material you are using?   9. _____

10. After spot welding using one-half the correct spacing, measure the diameter of each spot weld and record the size.

    1. _____
    2. _____
    3. _____
    4. _____
    5. _____
    6. _____
    7. _____
    8. _____
    9. _____
    10. _____
    11. _____
    12. _____

11. Does the weld size increase, decrease, or remain the same as each weld is made? _____

_____

INSPECTION
Perform a peel test on your finished welds. See Job 19-1 for instructions.

Instructor's initials: _____

Lesson 20A

ARC-RELATED WELDING PROCESSES

Name _____ Date _____ Lesson Grade _____
Class _____ Instructor _____

LEARNING OBJECTIVE
★ You will be able to describe special welding processes.

INSTRUCTIONS
Carefully read Headings 20.1 through 20.8 of the text. Also study Figures 20-1 through 18-26 in the text. Then complete the following problems.

1. *True or False?* A welding process is any method that can produce intermingling of the molecules.   1. _____

2. Match the abbreviations at the left with the welding processes at the right.

   _____ TW   A. Friction welding
   _____ SW   B. Forge welding
   _____ CW   C. Laser beam welding
   _____ FRW   D. Submerged arc welding
   _____ USW   E. Ultrasonic welding
   _____ LBW   F. Cold welding
                     G. Thermit welding
                     H. Arc stud welding
                     I. Coextrusion welding

3. Name the parts indicated on the drawing below.

   A. _____
   B. _____
   C. _____
   D. _____
   E. _____
   F. _____
   G. _____
   H. _____

4. The welding operator must set the _____ when automatic submerged arc welding.
   A. voltage
   B. electrode stickout
   C. wire feed speed
   D. travel speed
   E. alignment of the metal to be welded
   F. All of the above.

   4. _____

5. When multiple electrodes are used for submerged arc welding, which electrode combination is usually used?
   A. Parallel connection transverse position.
   B. Series connection transverse position.
   C. Multiple power connection tandem position.
   D. Multiple power connection transverse position.

   5. _____

6. Electroslag and electrogas welds are always done in a _____ welding position.

   6. _____

7. During electroslag welding, a layer of _____ _____ covers the molten weld metal.

   7. _____

8. Identify the parts of the welding process below.
   A. _____
   B. _____
   C. _____
   D. _____
   E. _____
   F. _____
   G. _____
   H. _____
   I. _____
   J. _____
   K. _____
   L. _____

9. What welding process is used to fill the weld joint when narrow gap welding? _____

10. Which of the following is the most significant problem with narrow gap welding?
    A. Shielding gas does not protect the molten metal.
    B. Thick sections cannot be welded.
    C. The arc must be directed to both sides of the joint.
    D. The guide tube is consumed during the welding process.

10. _____

11. *True or False?* Studs and fasteners can be welded to aluminum.

11. _____

12. When arc stud welding, a _____ _____ contains the molten metal during arcing.

12. _____

13. Label the parts of the arc stud welding gun.

    A. _____
    B. _____
    C. _____
    D. _____

14. *True or False?* Studs have a small nib or bump in them which always contain a small amount of flux.

14. _____

15. The orifice gas in PAW cannot contain _____.
    A. oxygen
    B. nitrogen
    C. helium
    D. argon

15. _____

16. *True or False?* The transferred arc can only weld on materials that conduct electricity.

16. _____

17. What current and arc voltage should be used to weld 0.125″ (32mm) thick stainless steel using PAW?

    current: _____
    voltage: _____

18. When does a plasma torch produce the best results? _____
_____

19. Label the parts of the figure below.

   A. _____
   B. _____
   C. _____
   D. _____
   E. _____
   F. _____

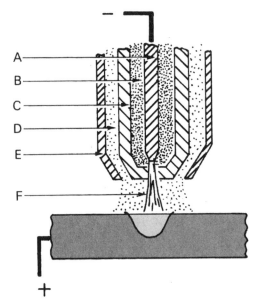

20. *True or False?* The plasma arc welding circuit shown above is a nontransferred arc.

20. _____

Lesson 20B

SOLID-STATE AND OTHER WELDING PROCESSES

Name _____ Date _____ Lesson Grade _____
Class _____ Instructor _____

LEARNING OBJECTIVE
★ You will be able to describe solid-state welding processes.

INSTRUCTIONS
Carefully read Headings 20.9 through 20.21 of the text. Also study Figure 20-27 through 20-61 in the text. Then complete the following problems.

1. *True or False?* During solid-state welding, the temperature of the material being welded is less than the melting point.

    1. _____

2. Match the following abbreviations with the correct name.

    _____ EXW   A. Ultrasonic welding
    _____ LBW   B. Diffusion welding
    _____ USW   C. Thermit welding
    _____ DFW   D. Forge welding
    _____ CEW   E. Friction welding
    _____ TW   F. Explosion welding
    _____ CW   G. Coextrusion welding
    _____ FRW   H. Laser beam welding
                             I. Cold welding

3. Cold welding is used on _____ metal.
    A. hard
    B. ductile
    C. brittle
    D. thick

    3. _____

4. How are metals to be cold-welded prepared? _____
    _____

5. *True or False?* Forge welding is usually limited to the welding of solid steel stock.

    5. _____

6. Parts to be forge-welded are heated to a _____ heat in a _____ forge.

    6. _____
    _____

7. *True or False?* In friction welding, heat is generated when one part rotates against a stationary (not rotating) part.

7. _____

8. It takes _____ to perform a friction weld.

8. _____

9. *True or False?* In inertia welding, both parts to be welded are rotated using flywheels rotating at high speeds.

9. _____

10. *True or False?* Ultrasonic welding equipment converts high-frequency electrical power into mechanical vibrations.

10. _____

11. List four advantages of ultrasonic welding. _____

_____

_____

_____

12. Which element is more chemically active: aluminum or iron?

12. _____

13. After the thermit mixture is ignited, a temperature of _____ °F (_____ °C) will be reached.

13. _____

14. Why is the thermit welding process considered to be safe? _____

_____

_____

15. High-energy-density welding produces welds that are _____ and _____.

15. _____

_____

16. Label the parts of the electron beam welder shown below.

A. _____

B. _____

C. _____

D. _____

E. _____

F. _____

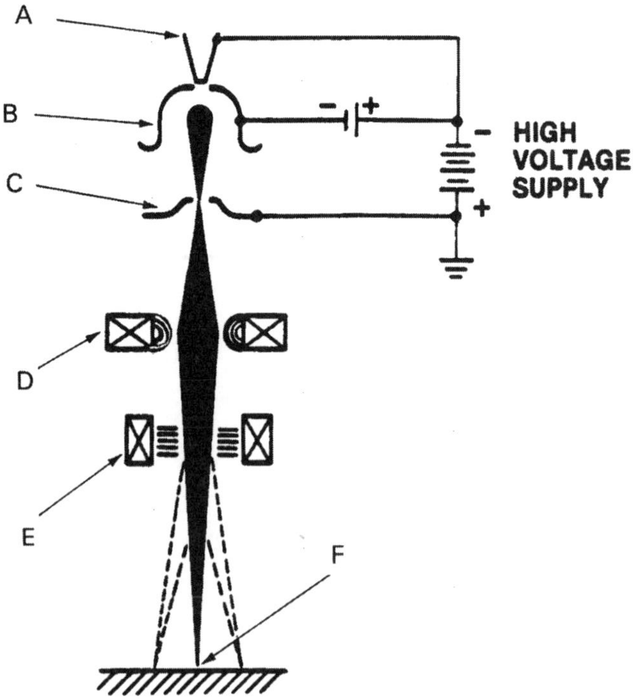

17. A complete penetration electron beam weld on _____ (_____mm) thick steel plate, can be made in one pass.

17. _____

18. In electron beam welding, the voltage difference between the emitter and the anode is known as the _____.
    A. primary voltage
    B. vacuum voltage
    C. accelerating voltage
    D. secondary voltage

18. _____

19. A(n) _____ laser can be used to weld continuously.
    A. ruby
    B. Nd-YAG
    C. CO_2
    D. All of the above.

19. _____

20. List the parts of the CO_2 laser shown below.

    A. _____
    B. _____
    C. _____
    D. _____
    E. _____
    F. _____
    G. _____
    H. _____

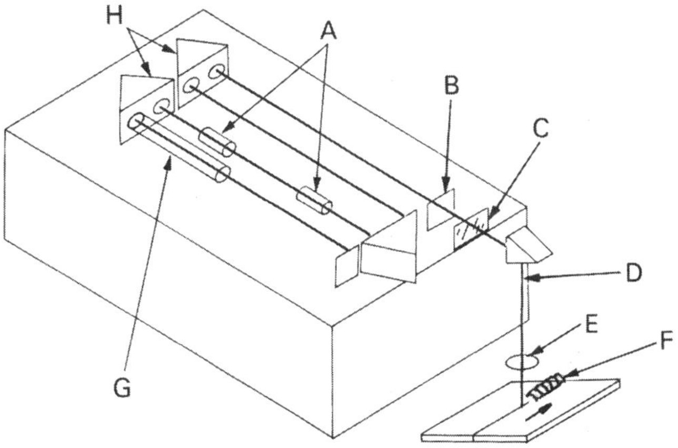

Lesson 21

FERROUS WELDING APPLICATIONS

Name _____ Date _____ Lesson Grade _____
Class _____ Instructor _____

LEARNING OBJECTIVE
★ You will be able to weld stainless steel, cast iron, and other ferrous alloys.

INSTRUCTIONS
Carefully read Headings 21.1 through 21.11 of the text. Also study Figures 21-1 through 21-18 in the text. Then complete the following problems.

1. _____ percent alloying elements are present in a low-alloy steel.   1. _____

2. _____ is not recommended for welding low-alloy steels.   2. _____
    A. SMAW
    B. OFW
    C. GMAW
    D. FCAW

3. List the *four* general classifications of stainless steel. _____
   _____
   _____
   _____

4. _____ is added to steel to make it a stainless steel.   4. _____

5. The "L" after an alloy designation (i.e. 316L) means _____.   5. _____

6. Which of the following statements are *true* for martensitic stainless steel?   6. _____
    A. It has good strength and is corrosion-resistant at elevated temperatures.
    B. It is often welded using an austenitic (300 series) filler metal.
    C. It contains 15%-20% chromium.
    D. Postweld heat treatment is required.

7. Does chromium martensitic steel require preheating prior to welding?   7. _____

303

8. List a stainless steel electrode that meets each of the following descriptions.

   A. SMAW electrode

   B. Low-hydrogen SMAW electrode

   C. GMAW electrode

   D. Low-carbon GMAW electrode

   8. A. _____

   B. _____

   C. _____

   D. _____

9. What type of shielding gases may be used to GMAW stainless steel? _____
   _____
   _____

10. One problem that occurs when welding stainless steel is grain growth. List three ways to reduce the problem.
    _____
    _____
    _____

11. *True or False?* To prevent porosity when welding stainless steel, a long arc length and a backing gas should be used.

    11. _____

12. *True or False?* Nickel-based alloys are used for high-temperature applications.

    12. _____

13. The ultimate tensile strength of a maraging steel can reach _____ KSI or _____ MPa.

    13. _____

14. Maraging steel contains _____% carbon maximum, _____% nickel, and _____% cobalt.

    14. _____

15. List the four principal types of cast iron. _____
    _____
    _____
    _____

16. _____ is not recommended for joining cast iron.
    A. Braze welding
    B. SMAW
    C. GMAW
    D. OFW

    16. _____

17. Why is cast iron preheated prior to welding? _____
    _____

18. A. What preheat temperature should be used on cast iron when SMAW?

    B. What preheat temperature should be used on cast iron when oxyfuel gas welding?

    18. A. _____

    B. _____

19. How is flux added to the weld when oxyfuel gas welding? _____
    _____
    _____

20. What post-weld heat treatment is recommended for precipitation-hardening stainless steel? _____

_____

_____

Job 21-1

LEARNING OBJECTIVE

★ In this job, you will learn to make a square-groove weld on a butt joint on stainless steel in the flat welding position. The process used will be shielded metal arc welding (SMAW).

Name _____
Date _____ Class _____
Instructor _____
Grade _____

1. Obtain eight pieces of 304, 304L, 308, or 308L stainless steel. Other types may also be used. Each piece should measure 6″ × 2″ × 3/16″ (152mm × 51mm × 4.8mm). Also obtain four 1/8″ (3.2mm) diameter E308 or 308L electrodes. Other electrodes may also be used. Your instructor will recommend other electrodes that may be used.

2. Use a stainless steel wire brush to remove the oxides from the stainless steel plates. Clean along the 6″ (152mm) edges.

3. Set the machine for DCEP (DCRP). The correct current for each electrode can be obtained from the manufacturer. Figure 12-19 in the text can also be used; however, the amperage should be set at 10%–20% less than that shown in Figure 12-19. Use the E6010 column for the 308 and 308L electrodes.

4. What current should be used for a 1/8″ (3.2mm) 308 or 308L electrode?   4. _____

5. Align two pieces of stainless steel to form a 6″ (152mm) long square-groove butt joint. Leave a 1/16″ to 1/8″ (1.6mm to 3.2mm) root gap.

6. Tack weld each end and the middle of the joint.

7. Weld the joint in the flat welding position. Use the keyhole method to obtain complete penetration. Maintain a short arc length to prevent porosity and spatter.

8. Chip the slag, brush with a stainless steel wire brush, and examine the bead. Refer to Figure 12-25 in the text.

9. Make any corrections required, and weld three additional butt welds. You may want to attempt the final butt weld as a horizontal or vertical weld.

INSPECTION

The beads should be straight and have complete penetration. The beads should have a uniform buildup, evenly spaced ripples, and good fusion with no overlap or undercut. Save these welded samples for use in Job 21-2.

Instructor's initials: _____

Job 21-2

LEARNING OBJECTIVE	Name _____
★ In this job, you will learn to make a fillet weld on a stainless steel T-joint in the flat position.	Date _____ Class _____ Instructor _____ Grade _____

1. Obtain four 1/8″ (3.2mm) diameter E308 or E308L electrodes and the samples that were welded in Job 21-1.

2. Prior to welding, the plates must be cleaned again. Stainless steel wire brush an area down the length of two of the welded samples. Wire brush the edge of two other welded samples.

3. Align one clean edge onto the clean area of another plate to form a T-joint. Tack weld the ends and middle of the plates.

4. Weld a fillet weld in the flat or horizontal welding position.

5. Remove the slag and examine the weld. If the weld does not appear correct, discuss your technique with your instructor.

6. Weld the opposite side of the joint you have already welded. Then weld the other two plates together using a double fillet weld.

INSPECTION
Fillet welds in stainless steel should be very similar to fillet welds in mild steel. The surface of the bead should have regularly spaced ripples and an even buildup with no overlap or undercut.

Instructor's initials: _____

Job 21-3

LEARNING OBJECTIVE

★ In this job, you will learn to make square-groove welds on a butt joint and fillet welds on a T-joint. The metal used will be stainless steel; the process used will be GTAW.

Name _____
Date _____ Class _____
Instructor _____
Grade _____

1. Obtain nine pieces of 304, 304L, 308, 308L, or another type as your instructor advises. Each piece should be 6" × 2" × 1/8" (152mm × 51mm × 3.2mm). Obtain three 36" (92cm) lengths of 3/32" (2.4mm) diameter filler rod, ER308, ER308L, or another alloy as your instructor advises.

2. Use a 3/32" (2.4mm) diameter thoriated tungsten electrode. Sharpen the electrode as discussed in Heading 14.3.9 and Figures 14-29 and 14-30 in the text.

3. A. What amperage should be used to butt weld 1/8" (3.2mm) stainless steel? (Refer to Figure 14-21 in the text.)   3. A. _____

   B. What amperage should be used to fillet weld 1/8" (3.2mm) stainless steel?   B. _____

4. Set the machine for DCEN (DCSP). Set the amperage on the machine. Also, adjust the argon flow rate per Figure 14-21 in the text.

5. Clean the edges of the material to be welded, using a stainless steel wire brush.

6. Align two pieces to form a 6" (152mm) square-groove butt weld. Leave a 1/16" (1.6mm) gap. Tack weld the middle and both ends of the joint.

7. Weld the joint using the keyhole method. Full penetration can be achieved in one pass.

8. Examine the completed joint and discuss your results and techniques with your instructor.

9. Align and tack another piece to the pieces already welded. Weld the butt weld complete.

10. Follow the figure below to complete a welded assembly. Use the two butt welds made previously in this weldment. Note that one of the fillet welds has intermittent welds. This is to allow you to practice your starts and stops. If you are having any difficulty while welding this assembly, contact your instructor to discuss the techniques required for welding stainless steel.

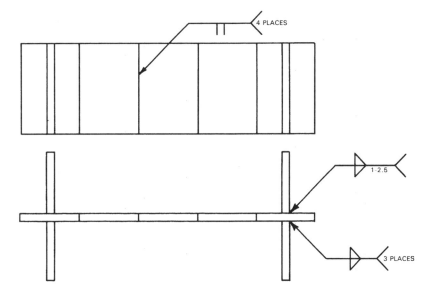

INSPECTION
Each fillet weld must be even in width, with evenly spaced ripples in the beads. The beads should be convex with no overlapping or undercutting. Each butt weld should have evenly spaced ripples, and a constant width and contour. It should also have full penetration over the entire length of the weld.

Instructor's initials: _____

Job 21-4

LEARNING OBJECTIVE
* In this job, you will learn to butt weld cast iron using the SMAW process.

Name _____
Date _____ Class _____
Instructor _____
Grade _____

1. Obtain six pieces of cast iron measuring approximately 6″ × 2″ × 3/8″ (152mm × 51mm × 9.5mm).

Note: Any available size will do.

2. Prepare the pieces for butt welding as described in Heading 21.10 and Figure 21-17 in the text. Prepare both edges of each piece, so that both edges can be welded. Bevel any pieces over 1/4″ (6.4mm) in thickness.

3. List three electrode types used to weld cast iron. _____
_____
_____
_____

4. Obtain six 5/32″ (4mm) diameter electrodes used for welding 3/8″ (9.5mm) thick cast iron.

5. Set the machine for DCEP (DCRP) or ac. Follow the electrode manufacturer's suggested current settings. The current used for welding cast iron is the same or slightly less than the current used for the same thickness of mild steel.

6. Preheat the cast iron to 500°F–600°F (260°C–316°C). Place the pieces in a preheated oven for one-half hour. If an oven is not available, use an oxyfuel gas torch and temperature-indicating crayons or liquid, to determine the correct preheat temperature.

7. Align two pieces to form a 6″ (152mm) long, V-groove butt weld. Leave a 1/16″ to 1/8″ (1.6mm to 3.2mm) root gap. Tack weld the pieces at each end and in the middle.

8. Weld the butt weld. The welding is similar to welding mild steel. Use stringer beads and clean the slag between each bead.

9. Weld additional pieces of cast iron on to the initial two pieces until five V-groove butt welds are completed. Follow steps 6, 7, and 8.

INSPECTION
Inspect the weld as you would a mild steel weld. The surface of the bead should have an even buildup and regularly spaced ripples. The root should have consistent penetration.

Instructor's initials: _____

Lesson 22

NONFERROUS WELDING APPLICATIONS

Name _____ Date _____ Lesson Grade _____
Class _____ Instructor _____

> **LEARNING OBJECTIVE**
> ✶ You will be able to weld aluminum, copper, titanium, and their alloys. You will also be able to describe other nonferrous applications and the welding of plastics.
>
> **INSTRUCTIONS**
> Carefully read Headings 22.1 through 22.12.1 of the text. Also study Figures 22-1 through 22-23 in the text. Then complete the following problems.

1. Which of the following are classified as nonferrous metals or alloys?
   A. Copper
   B. Aluminum
   C. Zirconium
   D. Cast iron
   E. Titanium
   F. Zinc
   G. Maraging steel

   1. _____

2. Match the aluminum alloy classification with its major alloying element.

   _____ 6061   A. Silicon
   _____ 5056   B. Magnesium and silicon
   _____ 1100   C. 99% aluminum (minimum)
   _____ 2024   D. Copper
   _____ 7075   E. Other
   _____ 4043   F. Zinc
                       G. Manganese
                       H. Magnesium

3. Which of the following is *not* a reason why aluminum is difficult to weld?
   A. Aluminum oxidizes at high temperatures.
   B. Aluminum oxide is heavier than aluminum.
   C. Aluminum reflects the heat of the arc.
   D. Aluminum melts before it changes color.

   3. _____

4. _____ current is recommended for GTAW aluminum. 4. _____

5. What type of tungsten electrode should be used when GTAW aluminum? _____
_____

6. In each of the following, select the best aluminum filler alloy for the alloy combination and criteria. Use Figure 22-2 in the text.

   A. Which filler alloy gives the best strength when welding 1100 aluminum to itself? 6. A. _____

   B. Which filler alloy gives the best results when welding 2219 to itself? B. _____

   C. Which filler alloy is easiest to weld with if welding 6061 to itself? C. _____

   D. Which filler alloy has the best ductility and ease of welding when joining 6061 to 5052? D. _____

   E. Which filler alloy has the best strength, ductility, ease of welding, and color match after anodizing, when welding 5052 to itself? E. _____

7. *True or False?* When welding thick section aluminum castings, the casting should be preheated to 400°F–500°F (204°C–260°C). 7. _____

8. _____ is *not* listed in Heading 22.5 of the text as a process used to join magnesium. 8. _____
   A. Brazing
   B. SMAW
   C. GTAW
   D. RSW

9. What may be used to control the fluid (liquid) weld metal when welding die castings? _____
_____

10. A(n) _____ flame should be used when oxyfuel gas welding die castings. 10. _____
    A. carburizing
    B. neutral
    C. oxidizing
    D. Any type of flame.

11. *True or False?* A very small tip should be used to oxyacetylene weld die castings. 11. _____

12. The metals that are alloyed to form bronze are _____ and _____. 12. _____
_____

13. *True or False?* Only oxygen-bearing copper can be fusion-welded easily. 13. _____

14. A. What type of current should be used to GTAW copper? _____
_____

    B. What type of tungsten electrode should be used? _____
_____

15. *True or False?* Copper has a high thermal conductivity so a preheat of 400°F–500°F (204°C–260°C) is often used.

15. _____

16. When arc welding brass, a _____ type electrode filler metal is used.

16. _____

17. Label the parts of the titanium welding set-up shown below.

   A. _____

   B. _____

   C. _____

   D. _____

   E. _____

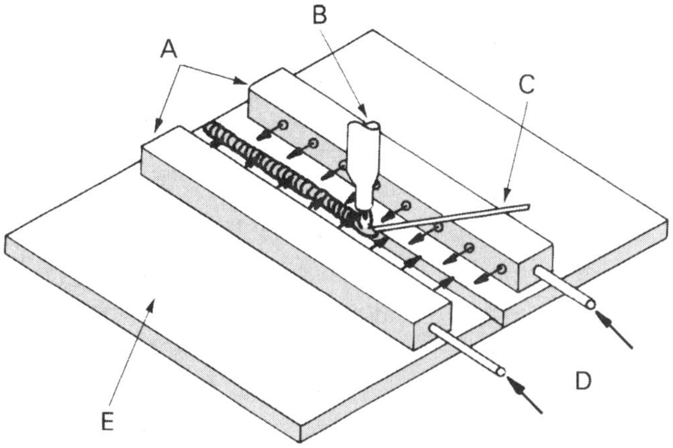

18. A _____ _____ must be worn whenever you enter a gas-filled area.

18. _____

19. What is the biggest problem with welding titanium and its alloys? _____

_____

_____

20. *True or False?* To increase the welding gas temperature (for welding plastic), the gas flow is increased.

20. _____

Job 22-1

LEARNING OBJECTIVE

★ In this job, you will practice butt welds and fillet welds on aluminum using the GTAW process. During this job assignment, you will make a welded assembly as shown in the figure at the end of this job.

Name _____
Date _____ Class _____
Instructor _____
Grade _____

1. Obtain 10 pieces of aluminum 6061, 5052, or 1100. Each piece should measure 6″ × 2″ × 1/8″ (152mm × 51mm × 3.2mm). Complete all welds shown in the drawing in #7. Follow the welding order outlined in this job assignment.

2. What filler metal is recommended for ease of welding for the base metal you will be welding? Refer to Figure 22-2 in the text.

    2. _____

3. What diameter electrode is recommended for 1/8″ (3.2mm) thick aluminum? Refer to Figure 14-20 in the text.

    3. _____

4. Obtain six 36″ (914mm) lengths of filler metal, 3/32″ (2.4mm) diameter. The type of filler metal is given in question #2. Also, obtain one pure tungsten electrode. The diameter is given in question #3. Install the electrode in the torch.

5. Clean the base metal by dipping the aluminum in a cleaning solution or by cleaning with a stainless steel brush.

6. Set the machine for ac welding. Adjust the current and the argon flow as described in Figure 14-19 in the text.

7. Complete the welds shown below.

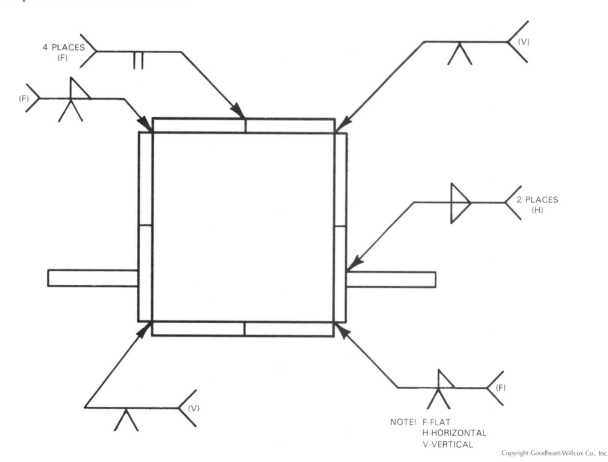

NOTE! F-FLAT
H-HORIZONTAL
V-VERTICAL

Copyright Goodheart-Willcox Co., Inc.

8. Align two pieces of aluminum to form a 6" (152mm) square-butt joint. Tack weld in the middle and on each end. A root opening of up to 1/16" (1.6mm) should be used.
9. Weld the butt weld complete.
10. Examine the butt weld and discuss the results with your instructor.
11. Use six pieces of aluminum to make three additional butt welds as described in #8.
12. Align one piece of aluminum on one of the butt-welded plates to form a T-joint. Tack weld in the middle and on each end. Weld a fillet weld down both sides of the T-joint.
13. Examine and discuss the results with your instructor.
14. Tack weld and weld the final piece of aluminum on a different butt-welded plate to form a T-joint.
15. Align two of the butt-welded plates to form a corner joint. Refer to the figure in #7. Tack weld in three places, then weld the outside corner joint in the flat weld position.
16. Discuss the results with your instructor.
17. Weld the inside of this corner joint.
18. Align the other two butt welded plates to form a corner joint. Weld the outside corner weld first, then the inside corner weld.
19. Align the two corner weld assemblies to form an enclosed box. Refer to the figure shown in #7. Tack weld the remaining corners of the box.
20. Weld the outside corner joints in the vertical welding position. Refer to Heading 14.10 and Figures 14-52, 14-53, and 14-54 in the text for vertical GTAW.

INSPECTION
Each weld should have beads of even width and evenly spaced ripples in the beads. There should be no evidence of overlapping or undercutting. All butt welds must have 100% penetration.

Instructor's initials: _____

Lesson 23

PIPE AND TUBE WELDING

Name _____ Date _____ Lesson Grade _____
Class _____ Instructor _____

LEARNING OBJECTIVE
★ You will be able to perform welds on pipe and tube.

INSTRUCTIONS
Carefully read Headings 23.1 through 23.13 of the text. Also study Figures 23-1 through 23-44 in the text. Then complete the following problems.

1. *True or False?* If a pipe is said to be 4″ in diameter, the outside of the pipe is greater than 4″.

    1. _____

2. _____ has the highest strength.
   A. Cast iron pipe
   B. Low-alloy seamed pipe
   C. Stainless steel seamless pipe
   D. Plastic pipe

    2. _____

3. *True or False?* Tubing is measured and ordered by its inside diameter (ID).

    3. _____

4. Name the dimensions on the figure below.

   A. _____
   B. _____
   C. _____
   D. _____

5. One difficulty encountered with all pipe and tube welding is the _____ or _____ of the pipe or tube after the welding is completed.

    5. _____
       _____

Copyright Goodheart-Willcox Co., Inc.

319

6. List three ways a pipe can be prepared (beveled) for welding. _____

_____

_____

_____

7. What are the two ways a vertical butt weld can be made? 7. _____

_____

8. Which of the two ways listed in question #7 would require fewer weld passes to complete? 8. _____

9. *True or False?* The root pass in a pipe weld is usually made with an E6010 or E7010 electrode. 9. _____

10. Label the different passes in the pipe weld shown below.

    A. _____

    B. _____

    C. _____

    D. _____

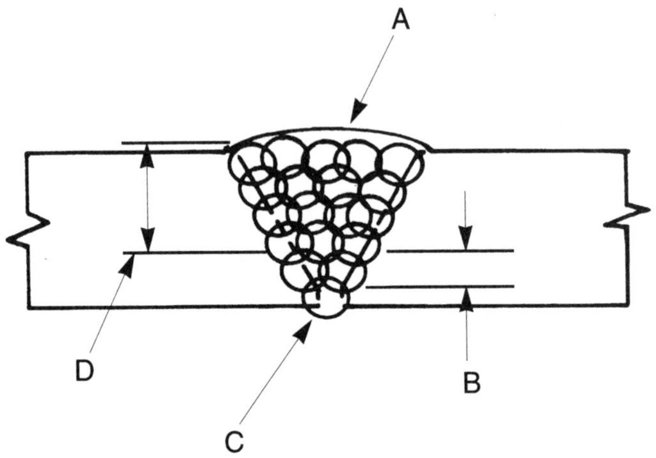

11. Other than E6010 and E7010, name a SMAW electrode commonly used for mild steel pipe welding. 11. _____

12. When using an E6010 electrode to weld an 8" (203mm) diameter pipe with a wall thickness of 3/16" (4.8mm), what electrode size and current should be used? Refer to Heading 12.5.1 and Figure 12-19 in the text.

    Electrode size: _____

    Current: _____

13. What electrode size and current should be used to weld the same pipe with an E7018 electrode? Refer to Heading 12.5.1 and Figure 12-20 in the text.

    Electrode size: _____

    Current: _____

14. What size welding tip orifice should be used to oxyacetylene weld a 2" (50.8mm) diameter mild steel tube with a wall thickness of 1/16" (1.6mm). Refer to Heading 6.5.3 and Figure 6-8 in the text. 14. _____

15. What oxygen and acetylene pressures should be used to weld the same tube (#14) using a positive-pressure type torch?

    Oxygen: _____

    Acetylene: _____

16. *True or False?* More than one welder can be used to weld large diameter pipe faster.

16. _____

17. The hot pass in pipe welds should be welded within _____ minutes after the root pass is completed.

17. _____

18. List two tests used to determine if there is a leak in a pipe or tube. _____
_____
_____

19. When welding pipes or tubes, they must be absolutely free of flammable _____ or _____ to prevent an explosion.

19. _____
_____

20. Label each of the positions shown with the correct AWS number and letter designation. Refer to Figure 31-5 in the text.

   A. _____
   B. _____
   C. _____
   D. _____

   A—MULTIPLE (NOT ROTATED)
   B—HORIZONTAL
   C—ROTATED FLAT
   D—MULTIPLE (NOT ROTATED)

Job 23-1

LEARNING OBJECTIVE ★ In this job, you will demonstrate your ability to weld a pipe joint in the 1G position using the SMAW process.	Name _____ Date _____ Class _____ Instructor _____ Grade _____

1. Obtain four sections of mild steel pipe, 3" (76.2mm) long, 4" (102mm) in diameter and about schedule #40. Other diameter pipe can be used. The pipe should be prepared for welding.

   What is the wall thickness of a schedule #40 pipe? _____

2. If the pipe is not prepared for welding, the edges may need to be beveled as shown in Figure 6-39 in the text.

3. Using an ac power source and an E6012 electrode, what electrode should be used for the wall thickness pipe you will be welding? 3. _____

4. What amperage range should be used for the electrode and wall thickness of the pipe? See Figure 12-19. 4. _____

5. Align two sections of pipe to form a V-groove butt joint. Leave a 1/16" (1.6mm) root gap.

6. Tack weld the pipe in at least three places.

7. Place the tack welded pipes into a weld positioner or use clamps to hold the pipes in the horizontal welding position.

8. Begin welding in the three o'clock position and weld to the twelve o'clock position. Refer to Figure 23-30 in the text for position of the electrode and the sequence of welding.

9. After welding from the three o'clock position to the twelve o'clock position, stop and rotate the pipe 90°. The section welded is now in the three to six o'clock position. Refer to Figure 23-30 in the text.

10. Begin welding in the three o'clock position where the last weld ended, and weld to the twelve o'clock position. Be careful to make a neat, strong blend with the previous bead. Refer to Heading 12.5.4.

11. Continue rotating the pipe and welding from the three o'clock to the twelve o'clock positions until the pipe is completely welded.

12. Depending on the wall thickness of the pipe you are welding, it may take more than one pass to complete the pipe weld. If a second weld pass is required, complete the root pass first. Then begin the second pass and continue to rotate and weld the pipe as described in steps 8 through 11.

13. Align another pipe section onto the two pieces already welded and tack weld it.

14. Follow the procedure described above and completely weld this section of pipe.

15. Weld the last section of pipe onto those already welded.

16. Shut down the arc welding station as described in Heading 12.5 of the text.

INSPECTION
Examine the weld bead. Each bead should be even in width and smooth. Each start and stop must have good fusion with the previous bead. There should not be any undercutting or overlapping. Examine the inside of the pipe. Penetration should be complete, but not excessive.

Instructor's initials: _____

Job 23-2

LEARNING OBJECTIVE
* In this job, you will demonstrate your ability to weld a pipe joint in the 2G and 5G positions using SMAW.

Name _____
Date _____ Class _____
Instructor _____
Grade _____

1. Obtain five sections of mild steel pipe 3″ (76.2mm) long, and 4″ (102mm) in diameter and about schedule #40. Other diameter pipe can be used. The pipe should be prepared for welding.

   What is the wall thickness of a 4″ diameter schedule #40 pipe? _____

2. If the pipe is not prepared for welding, the edges should be beveled as shown in Figure 23-10 in the text.

3. Using an ac power source and E6012 electrodes, what diameter electrode should be used for the wall thickness of your pipe?   3. _____

4. Using Figure 12-19, what amperage range should be used with your electrode?   4. _____

5. Align two sections of pipe to form a V-groove butt joint. Leave a 1/16″ (1.6mm) root gap.

6. Tack weld the pipe in at least three places.

7. Clamp the tack welded pipes into a weld positioner. There must be enough space to weld all around the pipe. This is the 5G position.

8. Begin welding in the six o'clock position, which is overhead welding. Weld from the six o'clock position to the 12 o'clock position, which will be uphill welding. Refer to Figure 23-28 in the text.

9. After reaching the 12 o'clock position, stop and restart at the 6 o'clock position. Weld uphill again to complete the pipe weld. *Do not rotate the pipe.*

10. If more than one pass is required to complete the pipe weld, weld the root pass completely before beginning the second pass or cover pass.

11. Tack weld another pipe section onto the pieces already welded.

12. Follow the procedure described above and completely weld this pipe joint.

13. Align another section of pipe to form a V-groove butt joint as described in #5 above and tack weld the joint.

14. Place the pipes into a weld positioner vertically, so that the joint is in the 2G position.

15. Weld the sections of pipe together. Refer to Heading 12-20 and Figure 12-51 in the text for the suggested electrode positions. Remember to point the electrode upward slightly to keep the weld pool from sagging.

16. Weld the joint completely, using one or more passes.

17. Weld the last section of pipe onto those already welded. Weld it in the 2G position.

18. Shut down the arc welding station. Refer to Heading 12.5.

INSPECTION
Examine the weld beads. The beads should be even in width and smooth. Each start and stop must have good fusion with a previous bead. There should not be any undercutting or overlapping. Examine the inside of the pipe. Penetration should be complete, but not excessive.

Instructor's initials: _____

Job 23-3

LEARNING OBJECTIVE
* In this job, you will demonstrate your ability to weld a pipe joint in the 5G position using the GTAW process.

Name _____
Date _____ Class _____
Instructor _____
Grade _____

1. Obtain four sections of mild steel pipe 3" (76.2mm) long, 4" (102mm) in diameter, and about a schedule #40. Other diameter pipe can be used. The pipe should be prepared for welding.

    What is the wall thickness of a 4" diameter schedule #40 pipe? _____

2. If the pipe is not prepared for welding, the edges should be beveled as shown in Figure 23-10 in the text.

3. What type electrode is recommended for welding mild steel? 3. _____

4. What size electrode and welding wire is recommended for welding the wall thickness of a schedule #40 pipe? 4. _____

5. What current range is suggested for welding the wall thickness pipe you will be welding? 5. _____

6. What type of current is recommended for the electrode you will be using? (Refer to Figures 12-17 and 12-18 in the text.) 6. _____

7. What type shielding gas and flow rate is recommended for mild steel? 7. _____

8. Set the machine for the correct current and current type. The electrode current is given in question #5 and the current type is given in question #6.

9. Align two sections of pipe to form a V-groove butt joint. Leave a 1/16" (1.6mm) root gap.

10. Tack weld the pipe in at least three places.

11. Clamp the tack welded pipes in the horizontal position. The weld joint is in the vertical position. There must be enough room to weld all around the pipe.

12. When pipe or tube welding, the circumference of the pipe or tube is divided like a clock face. The 12 o'clock position is at the top of the pipe or tube and 6 o'clock is at the bottom. Begin welding at the 5 o'clock position and weld uphill and over the top of the pipe to the 11 o'clock position. Clean the bead thoroughly. To complete the weld, begin at the 5 o'clock position and weld under the pipe and then uphill to the 11 o'clock position.

13. Depending on the thickness of the pipe you are welding, it may take more than one pass to complete the weld. If additional passes are required, repeat the procedure outlined in #12. Always clean the bead before starting the next pass.

14. Align another pipe section onto the two pieces already welded and tack weld it.

15. Follow the procedure described above and completely weld this section of pipe.

16. Weld the last section of pipe onto those already welded.

INSPECTION
Examine the weld beads. The beads should be even in width, smooth, and each start and stop must have good fusion with a previous bead. There should not be any undercutting or overlapping. Examine the inside of the pipe. Penetration should be complete, but not excessive.

Instructor's initials: _____

Job 23-4

LEARNING OBJECTIVE

★ In this job, you will demonstrate your ability to weld a pipe joint in the 2G position using the GMAW process.

Name _____

Date _____ Class _____

Instructor _____

Grade _____

1. Obtain four sections of mild steel pipe 3″ (76.2mm) long, and 4″ (102mm) in diameter and about schedule #40. Other diameter pipe can be used. The pipe should be prepared for welding.

   What is the wall thickness of a 4″ diameter schedule #40 pipe? _____

2. If the pipe is not prepared for welding, the edges should be beveled as shown in Figure 23-10 in the text.

3. Use a 0.035″ (0.89mm) diameter metal electrode wire and the short circuiting transfer method.

4. What is the recommended amperage (amperes) and voltage (volts) for welding the wall thickness of this pipe?

   4. _____

5. What type current is recommended for the electrode you will be using?

   5. _____

6. What type shielding gas and flow rate is recommended for mild steel?

   6. _____

7. Set the machine for the correct current and current type. See #4 and #5 above.

8. Align two sections of pipe to form a V-groove butt joint. Leave a 1/16″ (1.6mm) root gap.

9. Tack weld the pipe in at least three places.

10. Clamp the tack welded pipes into a weld positioner in the vertical position. The weld joint is in the horizontal welding position. There must be enough room to weld all around the pipe.

11. Begin welding in any location. Refer to Heading 12.7.6 and Figure 12-51 in the text for horizontal welding techniques. Remember, the electrode must be tipped 20° upward to counteract the sag of the molten metal. Weld completely around the pipe. Whenever you stop and restart, make sure you remove all slag.

12. Depending on the wall thickness of the pipe you are welding, it may take more than one pass to complete the pipe weld. If a second weld pass is required, complete the root pass first. Remove all slag. Then begin the second pass.

13. Align another pipe section onto the two pieces already welded and tack weld it.

14. Follow the procedure described above and completely weld this section of pipe.

15. Tack weld and weld the last section of pipe onto those already welded.

INSPECTION

Examine the weld beads. The beads should be even in width and smooth. Each start and stop must have good fusion with a previous bead. There should not be any undercutting or overlapping. Examine the inside of the pipe. Penetration should be complete, but not excessive.

Instructor's initials: _____

Lesson 23 Pipe and Tube Welding 327

Job 23-5

LEARNING OBJECTIVE
★ In this job, you will demonstrate your ability to weld a pipe joint in the 6G position using the GMAW process.

Name _____
Date _____ Class _____
Instructor _____
Grade _____

NOTE TO THE STUDENT

Welding in the 6G position is the most difficult position for any pipe weld. Before attempting to weld in the 6G position, practice welding in the 2G and 5G positions until you can produce consistently high-quality welds. The 6G position is a combination of the 2G and 5G positions. Very few pipe welds on a construction site are made in this position. However, many pipe weld qualification tests are taken using this 6G position. You should practice this pipe weld position and become very good at it before applying for a pipe welding job.

1. Obtain four sections of mild steel pipe 3″ (76.2mm) long, and 4″ (102mm) in diameter and about schedule #40. Other diameter pipe can be used. The pipe should be prepared for welding.

    What is the wall thickness of the pipe you obtained?   1. _____

2. If the pipe is not prepared for welding, the edges should be beveled as shown in Figure 23-10 in the text.

3. Refer to Headings 15.9, 15.10, and 15.11 in the text for welding techniques.

4. Determine the following values required to weld this joint:

    Type of electrode   4. _____

    The best metal transfer method   _____

    Electrode diameter   _____

    Amperage range   _____

    Voltage range   _____

    Suggested shielding gas   _____

    Flow rate   _____

    Type of dc current   _____

5. Set the machine for the correct current and current type.

6. Align two sections of pipe to form a V-groove butt joint. Leave a 1/16″ (1.6mm) root gap.

7. Tack weld the pipe in at least three places.

8. Clamp the tack welded pipes at a 45° position in a weld positioner. The weld joint is also at a 45° angle. There must be enough room to weld all around the pipe.

9. Begin welding at the bottom of the pipe (the 6 o'clock position). Weld uphill to the top. As you move toward the top, the position and the angle of the electrode constantly changes. This is why the 6G position is so difficult. The electrode should be tipped 20° in the direction of travel. As you weld from the 8 o'clock to the 10 o'clock position, the electrode must also be tipped 20° upward to counteract the sag of the molten metal. When you reach the top of the pipe, stop. Restart at the bottom of the pipe and again weld uphill to the top. Whenever you stop and restart, make sure you remove all slag.

10. Depending on the wall thickness of the pipe you are welding, it may take more than one pass to complete the pipe weld. If a second weld pass is required, complete the root pass first. Remove all slag. Then begin the second pass.

Copyright Goodheart-Willcox Co., Inc.

11. Align another pipe section onto the two pieces already welded and tack weld it.

12. Follow the procedure described above and completely weld this section of pipe.

13. Tack weld and weld the last section of pipe onto those already welded.

INSPECTION
Examine the weld beads. The beads should be even in width and smooth. Each start and stop must have a good fusion with a previous bead. There should not be any undercutting or overlapping. Examine the inside of the pipe. Penetration should be complete, but not excessive. Producing high-quality pipe welds in the 6G position is difficult; but when done properly, there should be no difference between a pipe weld made in the 6G position and one made in any other position.

Instructor's initials: _____

Lesson 24

SPECIAL CUTTING PROCESSES

Name _____ Date _____ Lesson Grade _____
Class _____ Instructor _____

> **LEARNING OBJECTIVE**
> * You will be able to describe special cutting processes.
>
> **INSTRUCTIONS**
> Carefully read Headings 24.1 through 24.8 of the text. Also study Figures 24-1 through 24-19 in the text. Then complete the following problems.

1. Match the following processes with the correct abbreviations.

   A. _____ Laser beam cutting   FOC
   B. _____ Oxygen lance cutting   LOC
   C. _____ Flux cutting   AOC
   D. _____ Oxygen arc cutting   POC
   E. _____ Metal powder cutting   LBC

2. What must be worn during all cutting operations? _____
   _____

3. *True or False?* When metal powder cutting, oxygen should never be used to pressurize a metal powder container.   3. _____

4. Steel to be cut using LOC is heated to a _____ hot temperature.   4. _____

5. _____ people are required to perform an oxygen lance cutting operation.   5. _____

6. A _____ oxygen regulator is used with oxygen lance cutting.   6. _____
   A. single-stage nozzle type pressure
   B. single-stage stem type
   C. two-stage high volume
   D. All of the above.

7. _____ or _____ gas can be used for underwater cutting below 15′ (4.6m).   7. _____
   _____

Copyright Goodheart-Willcox Co., Inc.

329

8. *True or False?* Underwater cutting tips are larger than those used on the surface (land). 8. _____

9. List the four valves on an underwater cutting torch. _____
_____
_____
_____

10. Label the parts of the oxygen arc cutting torch shown below.

   A. _____
   B. _____
   C. _____
   D. _____
   E. _____

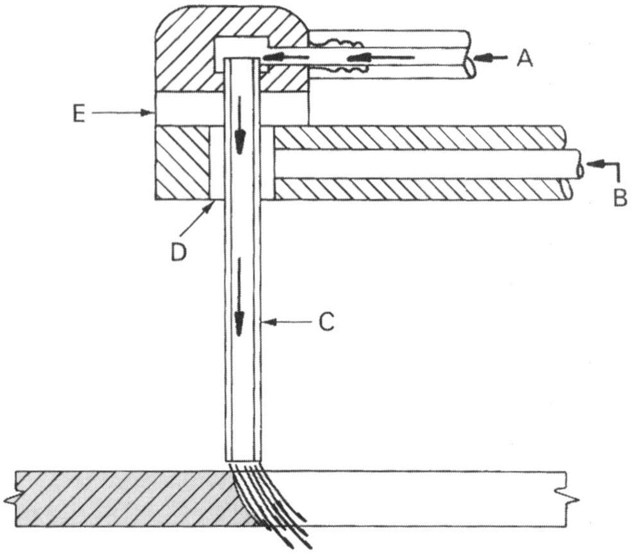

11. What type of electrode is often used when oxygen arc underwater cutting? _____
_____
_____

    What type current is used? _____

12. *True or False?* Chromium oxide is introduced into the cutting flame during metal powder cutting. 12. _____

13. *True or False?* Iron powder is used during metal powder cutting. When the iron powder is introduced through a separate orifice in the tip, there is a delay in the ignition of the iron powder. To prevent a delay in the cutting action, the metal is preheated. 13. _____

14. List the equipment required for metal powder cutting. _____
_____
_____
_____

15. *True or False?* A pressurized hopper produces a more constant and accurate powder flow than a vibrating type dispenser. 15. _____

16. Label the parts of the metal powder cutting lance shown below.

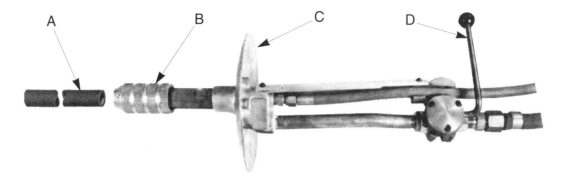

A. _____
B. _____
C. _____
D. _____

17. What is the function of the chemical fluxes in chemical flux cutting? _____
_____

18. During the chemical flux cutting process, where is the chemical flux added? _____
_____

19. Label the parts of the chemical flux cutting outfit shown below.

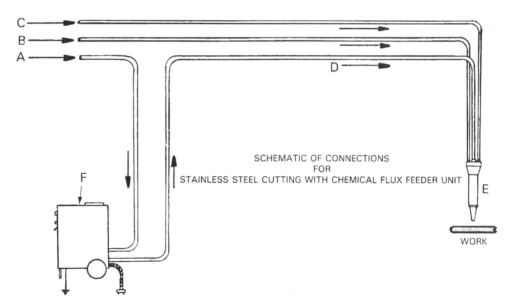

A. _____
B. _____
C. _____
D. _____
E. _____
F. _____

20. Which of the following is a *disadvantage* for laser beam cutting?
    A. The workpiece is not part of the electrical circuit.
    B. A laser beam creates a very narrow kerf.
    C. A laser beam is suitable for cutting metal up to approximately 1/2" (13mm) thick.
    D. A laser beam can be reflected.

Lesson 25

AUTOMATIC AND ROBOTIC WELDING

Name _____ Date _____ Lesson Grade _____
Class _____ Instructor _____

LEARNING OBJECTIVE
* You will be able to describe the principles of automatic welding and robot operation.

INSTRUCTIONS
Carefully read Headings 25.1 through 25.5 of the text. Also study Figures 25-1 through 25-26 in the text. Then complete the following problems.

1. _____ is *not* an advantage of automatic welding.   1. _____
    A. Increased productivity
    B. Increased labor
    C. Improved quality
    D. Reduced cost

2. List three feedback controls that a welder uses and state when each is used while turning on or turning off an oxyfuel gas torch. (These feedback controls can be sight, sound, smell, or touch.) _____
_____
_____
_____

3. A flow switch can be used to _____.   3. _____
    A. sound an alarm
    B. shut off the machine
    C. light a warning light
    D. All of the above.

4. *True or False?* Flow switches are a form of feedback control.   4. _____

5. In order for a solenoid gas valve to allow gas to flow, you must   5. _____
   _____.
    A. apply high gas pressure
    B. adjust the flow control knob
    C. apply the required voltage
    D. shut off the power

6. What does a solenoid-operated relay control? _____
_____

Copyright Goodheart-Willcox Co., Inc.

333

334   Modern Welding Lab Manual

7. *True or False?* One disadvantage of a relay is that a large voltage is required to control a smaller voltage.

7. _____

8. *True or False?* All automatic welding operations depend on the use of solenoid-operated valves.

8. _____

9. Label the parts of the GMAW circuit shown below.

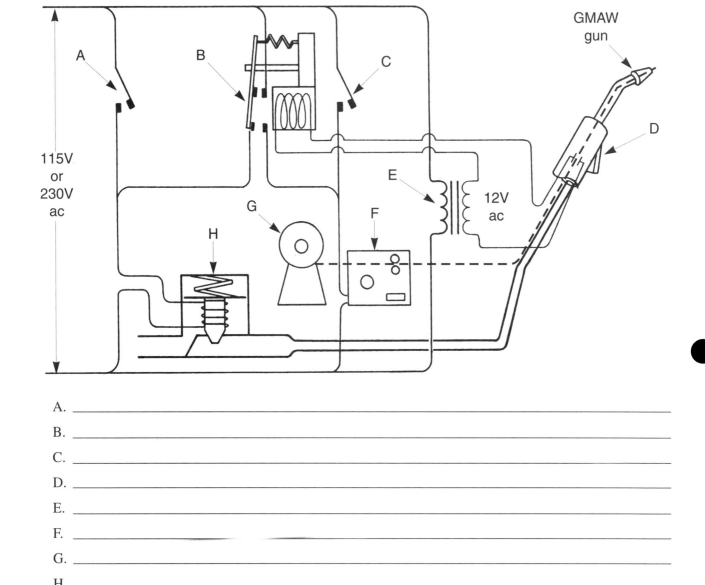

A. _____
B. _____
C. _____
D. _____
E. _____
F. _____
G. _____
H. _____

10. Describe what happens when the gun switch on a GMAW torch is pressed. _____
_____
_____
_____

11. The knob on a variable-speed motor is attached to a _____.
    A. variable resistor
    B. capacitor
    C. single-phase ac motor
    D. voltage regulator

11. _____

Copyright Goodheart-Willcox Co., Inc.

Name: _____ Lesson 25 Automatic and Robotic Welding 335

12. Label the five major parts of the robot system shown below.

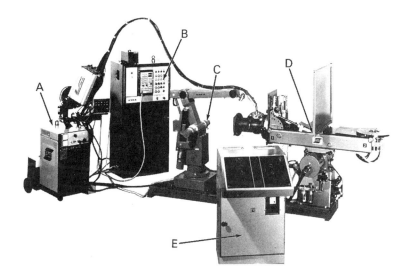

A. _____
B. _____
C. _____
D. _____
E. _____

13. Most robots have _____ axes.
    A. 4
    B. 5
    C. 6
    D. All of the above.

13. _____

14. Parts that are to be welded must be placed inside the robot's _____ _____.

14. _____

15. The "brain" of a robotic operation is called the _____. The responsibilities that it performs can be classified into three categories. List them below.

15. _____

_____
_____
_____

16. *True or False?* A robotic system uses electrical signals, solenoid operated valves, and solenoid operated relays to perform the program.

16. _____

17. List two advantages of using a positioner. _____
_____
_____

18. A. What button is pressed to make the robot execute a program?

18. A. _____

   B. What button is pressed if something goes wrong during the operation of the robot?

    B. _____

19. *True or False?* When programming a robot, the teach pendant has control over each axis of the robot and positioner.

19. _____

20. *True or False?* You should never enter the working volume of a robot when it is in operation.

20. _____

Lesson 26

METAL SURFACING

Name _____ Date _____ Lesson Grade _____
Class _____ Instructor _____

LEARNING OBJECTIVE
★ You will be able to perform metal surfacing.

INSTRUCTIONS
Carefully read Headings 26.1 through 26.10 of the text. Also study Figures 26-1 through 26-39 in the text. Then complete the following problems.

1. *True or False?* Hardfacing is when surfacing material is sprayed on at extremely high temperatures.

    1. _____

2. Which of the following is *not* an advantage of metal surfacing?
    A. Certain dimensions can be maintained under adverse conditions
    B. The service life of a part may be greatly increased
    C. More expensive alloys are required to fabricate the part
    D. Fewer replacement parts can be carried in stock

    2. _____

3. *True or False?* There are three surface treatments that give maximum resistance to corrosion, impact breaks, and abrasive wear at the same time.

    3. _____

4. A part that will be subjected to chemical wear should *not* have a(n) _____ applied.
    A. nickel-base alloy
    B. aluminum-base alloy
    C. cobalt-base alloy
    D. tungsten carbide mixture

    4. _____

5. Match the following hardness levels with the ability of a mill file to remove metal from the part being tested.

    _____ 60 Rockwell B
    _____ 60 Rockwell C
    _____ 400 Brinell
    _____ 15 Rockwell C

    A. Readily cut with moderate pressure
    B. Cannot be cut
    C. Removed easily
    D. Cut with great pressure only

6. A _____ _____ flame is used when surfacing with an oxyfuel gas process.

    6. _____

7. Match the following processes with the advantage of that process.

   _____ Oxyfuel gas
   _____ Thermal or arc spraying
   _____ Manual gas metal arc
   _____ Plasma arc
   _____ Electric arc spraying

   A. Any material can be bonded to any other
   B. Nonferrous metals are easily applied
   C. Most economical
   D. Easily used in various positions
   E. Flawless surface deposit

8. When hardfacing with an oxyfuel gas process, the tip should be _____ to _____ sizes larger than used with the same diameter rod when welding.

   8. _____

9. *True or False?* An oxyfuel gas torch can be used for flame spraying.

   9. _____

10. When surfacing with the shielded metal arc process, a _____ current is used. This is required so a _____ arc length can be maintained.

    10. _____

11. The recommended height of a deposit is _____ to 1/4, and the recommended width is 3/4 to _____. When laying a second bead, the beads should overlap 1/4 to _____ the width of the first bead.

    11. _____

12. Label the parts of the wire feed flame spraying torch shown below.

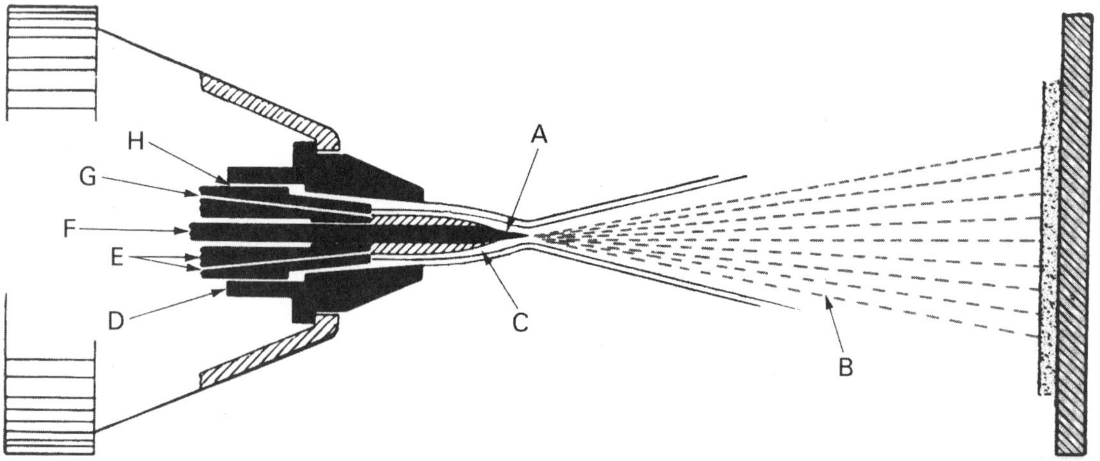

    A. _____
    B. _____
    C. _____
    D. _____
    E. _____
    F. _____
    G. _____
    H. _____

13. Material to be flame sprayed comes in different forms. Which of the following forms *cannot* be obtained?
    A. Liquid
    B. Powder
    C. Wire
    D. Powder placed inside a hollow metal wire or plastic tube

13. _____

14. Why is oxygen never used as a carrier gas in flame spraying? _____
_____

15. *True or False?* Ceramic material can be flame-sprayed.

15. _____

16. The tool shown below is referred to as a(n) _____ torch.

16. _____

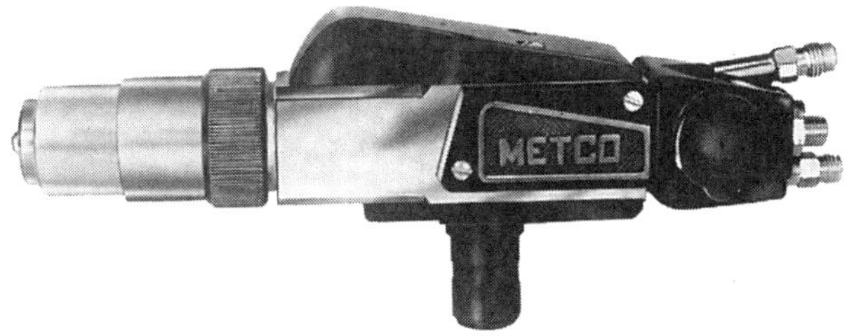

17. *True or False?* Material is flame sprayed in layers 0.001″ to 0.002″ (0.025mm to 0.050mm) thick.

17. _____

18. Electric arc spraying uses _____ current carrying wires that are automatically fed to the arc position in the gun. Arc spraying guns spray up to _____ lb. of wire per hour.

18. _____
_____

19. List three surface preparations used to prepare a metal for flame spraying. _____
_____
_____

20. If a defect is found in a part after surfacing, what should be done and how is it usually done? _____
_____
_____

Job 26-1

LEARNING OBJECTIVE
* In this job, you will demonstrate your ability to hardface a steel plate using an oxyfuel gas torch.

Name _____
Date _____ Class _____
Instructor _____
Grade _____

1. Obtain one piece of mild steel plate 6″ × 6″ × 3/16″ thick (152mm × 152mm × 4.76mm). Other thickness can be used.

2. What type of flame should be used? 2. _____

3. What size welding tip should be used for welding 3/16″ (4.76mm) plate? See Figure 6-8 in the text. 3. _____

   What size tip should be used to hardface? _____

4. What diameter welding rod should be used to hardface? (See Figure 6-8 in the text.) 4. _____

5. Obtain six hardfacing rods.

6. The plate will be surfaced as follows:

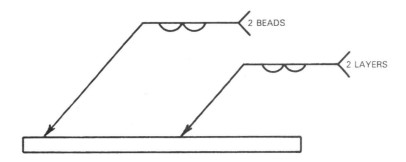

7. Light the torch and adjust it to the flame type listed in question #2.

8. Heat one edge of the metal with the torch to 2200°F (1204°C). Steel is a bright red-orange color at this temperature. Do not melt the base metal.

9. After the metal has been heated, touch the hardfacing rod to the steel.

10. Run a bead down one side of the plate as if you were welding, but do not melt the base metal. Keep the metal at the correct temperature. The base metal, not the torch, should melt the rod. The bead can be run using either the forehand or backhand method. (Refer to Figures 26-2 through 26-5 in the text.)

11. After completing one bead, move over 3/4″ (19mm) and run a second bead. The toes of the two beads should not touch.

12. How much should one bead overlap another? 12. _____

13. Now move into the center of the plate and run a bead down the center.

14. Run a second bead so that it overlaps the first bead by the amount listed in question #12.

15. Run a third and fourth bead on the steel so each overlaps the previous by the correct amount.

16. Now lay another bead so it is centered between two beads. In this way the hardfacing material can be built up. Refer to Figure 26-10 in the text.

17. Lay two more beads between the beads made in question #15. The buildup should be approximately 1/4″ (6.3mm).

INSPECTION
Examine the beads you have laid. Each bead should be even in width and uniform in height. There should be no visible cracks or chips in the beads. Any start and stop should have good fusion into the previous bead.

Instructor's initials: _____

Lesson 27

PRODUCTION OF METALS

Name _____ Date _____ Lesson Grade _____
Class _____ Instructor _____

> **LEARNING OBJECTIVE**
> ✱ You will be able to describe the production of many metals that have industrial applications.
>
> **INSTRUCTIONS**
> Carefully read Headings 27.1 through 27.8 of the text. Also study Figures 27-1 through 27-22 in the text. Then complete the following problems.

1. _____ and _____ are the most common metals in industrial use.   1. _____

2. _____ is added to iron to form straight carbon steel.   2. _____
    A. Carbon
    B. Chromium
    C. Manganese
    D. Nickel

3. *True or False?* The atmosphere inside a steelmaking furnace is an oxidizing atmosphere.   3. _____

4. List the materials required to operate a blast furnace. _____
   _____
   _____

5. List the five major functions of a blast furnace. _____
   _____
   _____
   _____
   _____

6. *True or False?* During the operation of the blast furnace, the melting temperature of the iron becomes lower as it combines with the excess carbon from the coke.   6. _____

7. _____ iron contains the least amount of carbon.
   A. Gray cast
   B. Malleable
   C. Ductile
   D. Wrought

7. _____

8. The _____ furnace(s) is/are *not* used for producing steel.
   A. open hearth
   B. blast
   C. electric
   D. cupola
   E. pudding

8. _____

9. Refer to Heading 27.2.1 of the text. What does B.O.F. mean? _____
   _____

10. A(n) _____ _____ is used to blow oxygen into the furnace during the basic oxygen process.

10. _____

11. The firebricks in an open hearth furnace are arranged in a _____ pattern.

11. _____

12. *True or False?* The firebricks in an open hearth furnace are used to preheat the incoming fuel gases.

12. _____

13. *True or False?* In an electric furnace, an arc is struck between flux cored electrodes and the metal in the furnace.

13. _____

14. Which furnace, the electric furnace or the crucible furnace, produces higher quality steel? _____
    _____
    _____

15. To control temperature of the metal in an induction furnace, the _____ and the _____ of the alternating current must be controlled.

15. _____
    _____

16. Absorbed gases in steel cause porosity and inclusions. The _____ furnace removes these gases.
    A. electric
    B. crucible
    C. induction
    D. vacuum

16. _____

17. Copper is first refined in a _____ furnace. It is further refined using an _____ process.

17. _____
    _____

18. The _____ furnace(s) is/are used to produce brass and bronze.
    A. cupola
    B. crucible
    C. electric
    D. open hearth

18. _____

19. The continuous casting process is used to eliminate some costly operations. Label the parts of the continuous casting process shown below.

A. _____
B. _____
C. _____
D. _____
E. _____
F. _____
G. _____
H. _____

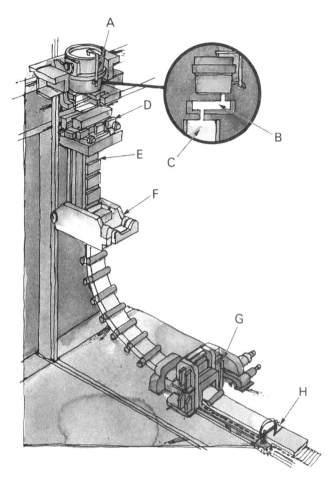

20. Aluminum is produced using an electrolysis process called the _____ Process. Label the parts of the electrolytic cell shown below.

20. _____

A. _____
B. _____
C. _____
D. _____
E. _____

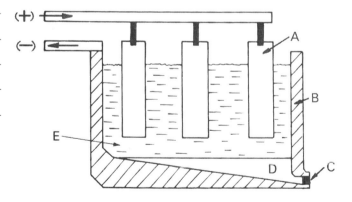

Lesson 28

METAL PROPERTIES AND IDENTIFICATION

Name _____ Date _____ Lesson Grade _____
Class _____ Instructor _____

> **LEARNING OBJECTIVE**
> ✱ You will be able to describe the properties of different metals and identify them.
>
> **INSTRUCTIONS**
> Carefully read Headings 28.1 through 28.11 of the text. Also study Figures 28-1 through 28-16 in the text. Then complete the following problems.

1. Match the following carbon contents with the correct designation for that carbon range.

    _____ 0.30% to 0.55% carbon   A. Very high-carbon steel
    _____ 0.003% carbon   B. Low-carbon steel
    _____ less than 0.030% carbon   C. Cast iron
    _____ 1.8% to 4.0% carbon   D. Medium-carbon steel
    _____ 0.55% to 0.80% carbon   E. Wrought iron
                                     F. High-carbon steel

2. The physical properties of iron and steel are greatly affected by four factors. Name them. _____
   _____
   _____
   _____
   _____

3. Some physical properties are listed below. Match each physical property with the correct description.

    _____ Compressive strength   A. How much a metal will stretch
    _____ Ductility   B. Resistance to being pulled apart
    _____ Toughness   C. Ease of fracturing
    _____ Tensile strength   D. Ability to be stretched
    _____ Brittleness   E. Resistance to penetration
    _____ Elongation   F. Prevents a crack from growing
    _____ Hardness   G. Withstands a squeezing force

Copyright Goodheart-Willcox Co., Inc.

4. *True or False?* A metal that has over 5% elongation is considered to be ductile, a metal that has less than 5% elongation is considered to be brittle.

4. _____

5. The following elements are often added to plain carbon steel to produce alloy steels. Each element will alter the steel's properties. Match the results of each of the following elements on steel.

    _____ Chromium
    _____ Nickel
    _____ Molybdenum
    _____ Vanadium
    _____ Tungsten
    _____ Manganese

    A. Increases toughness, increases strength at high temperatures
    B. Produces dense fine grains
    C. Increases resistance to corrosion
    D. Increases strength and responsiveness to heat treatment
    E. Retards grain growth
    F. Increases strength, ductility and toughness

6. Give a brief definition of these terms: (Refer to Heading 9.2 of the text for additional information.)

    Solidus: _____

    _____

    Liquidus: _____

    _____

7. Label the parts of the iron-carbon diagram shown below. Record your answers at the top of the following page.

    Note: The answers to F and G are *not* hypereutectoid and hypoeutectoid steels.

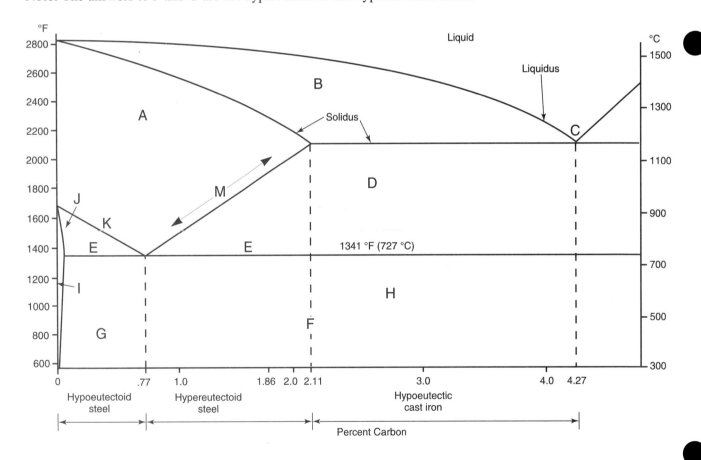

A. _____
B. _____
C. _____
D. _____
E. _____
F. _____
G. _____
H. _____
I. _____
J. _____
K. _____
L. _____
M. _____

8. *True or False?* Ferrite contains very small amounts of carbon, has a face centered structure, and is both ductile and tough.

8. _____

9. *True or False?* Cementite contains 6.69% carbon, is hard and brittle.

9. _____

10. Pearlite is a combination of _____ and _____. Pearlite contains _____% carbon.

10. _____

11. At the eutectic point, solid austenite transforms into two new solids. Name them. _____

12. The _____ is *not* a characteristic of the spark in a spark test.
    A. color of spark
    B. color of explosions
    C. number of explosions
    D. shape of explosions

12. _____

13. Which of the following results from an oxyacetylene torch test would indicate a metal was not suitable for welding?
    A. The weld pool is fluid and has good surface tension.
    B. The metal has an even, shiny appearance after solidification.
    C. The metal gives off numerous sparks.
    D. The metal is not rough and broken up into pits or porous spots.

13. _____

14. List four additional tests used to identify metals. Also give a brief explanation of how the test is done or the results of the test. _____

_____

_____

_____

_____

_____

_____

_____

15. Which of the statements below is *not* true for nonferrous metals?   15. _____
    A. They have distinctive colors.
    B. They are magnetic.
    C. They are usually relatively soft.
    D. They do not spark when touched to a grinding wheel.

16. Which of the nonferrous metals in Figure 28-15 in the text has the   16. _____
    highest melting point?

17. Copper, like other nonferrous metals, exhibits hot shortness. What are the characteristics of hot shortness?

_____

_____

18. A. What color is copper?   18. A. _____

    B. What is the melting temperature of copper?   B. _____

    C. Is copper magnetic?   C. _____

19. *True or False?* Aluminum has good electrical conductivity, heat   19. _____
    conductivity, and resistance to corrosion. It is lightweight, has
    excellent strength, has a gray color, and has a melting temperature
    of 1220°F (660°C).

20. Titanium is _____% heavier than aluminum and 40% lighter than   20. _____
    steel. It retains its strength up to _____°F (_____°C). Titanium is
    alloyed with _____ to increase its strength.   _____

Job 28-1

LEARNING OBJECTIVE
* In this job, you will use different tests to identify various metals.

Name _____
Date _____ Class _____
Instructor _____
Grade _____

1. List eight tests that can be used to identify metals. _____
   _____
   _____
   _____
   _____
   _____
   _____
   _____
   _____
   _____

2. Your instructor will supply a box containing ten or more different metals. The metals may include low-carbon steel, high-carbon steel, tool steel, stainless steel, or cast iron. The box may include copper, aluminum, magnesium, lead, or other metals.

3. Take one of the metals from the box.

4. Perform as many tests on it as required to determine what kind of material it is. Use the color and density test to narrow down the possibilities. Use the spark test to determine carbon and alloy content. **Caution:** Always use approved eye protection when grinding.

5. Make a list of the test(s) you used to determine the metal type.

6. When you have determined the metal type, write it down. If it is steel, also write down the approximate carbon content and any alloys it contains.

7. Select another piece of metal from the box and determine what type of metal it is. Write down the test(s) you used and what metal type it is.

8. Practice on as many different pieces as possible. When you feel confident of your abilities to identify different types of metal, contact your instructor.

9. Your instructor will select two different pieces of metal.

10. You will be asked to identify both pieces of metal. You will also be asked to tell which tests you used and why. Fill in the tests you used and the results on the following page.

METAL #1

Test Used	Test Results
_____	_____
_____	_____
_____	_____
_____	_____

Metal #1
identified as

_____

METAL #2

Test Used	Test Results
_____	_____
_____	_____
_____	_____
_____	_____

Metal #2
identified as

_____

Instructor's initials: _____

Lesson 29

HEAT TREATMENT OF METALS

Name _____ Date _____ Lesson Grade _____
Class _____ Instructor _____

> **LEARNING OBJECTIVE**
> ✶ You will be able to explain how various heat treatments are used to alter the properties of many metals.
>
> **INSTRUCTIONS**
> Carefully read Headings 29.1 through 29.16 of the text. Also study Figures 29-1 through 29-30 in the text. Then complete the following problems.

1. Name the three different times that heating can occur in the welding cycle. _____
   _____
   _____
   _____

2. The _____ is *not* an important factor to consider when heat treating.
   A. temperature to which a metal is heated
   B. length of time that the metal remains in the cooling medium
   C. rate at which the metal is cooled
   D. material surrounding the metal when it is heated

   2. _____

3. There are many ways used to measure temperatures during heat treatment. Name the tool shown below.

   3. _____

4. List three methods, other than flame heating, used to heat a part. _____
   _____
   _____
   _____

Copyright Goodheart-Willcox Co., Inc.

5. *True or False?* Cooling takes longer if a colder cooling medium is used.

5. _____

6. Heat loss takes place in a combination of three ways. Name all three. _____

_____

_____

_____

7. A steel that contains large amounts of _____ cannot be hardened. It cannot be hardened because it is the amount of _____ that determines the hardness.

7. _____

_____

8. When a steel is slow-cooled, _____ will form. Martensite is formed when a steel is _____.

8. _____

_____

9. Describe the physical characteristics of the following.

Bainite: _____

_____

_____

Pearlite: _____

_____

_____

Martensite: _____

_____

_____

10. After a steel is welded, there are two regions of grain sizes. The grain _____ region is next to the weld metal. The grain _____ region is next to the base metal.

10. _____

_____

11. The smallest grains are produced when a steel is heated to the _____ critical temperature, and then cooled.

11. _____

12. *True or False?* Since grain growth is a function of time and temperature, a higher temperature and a longer time will produce a larger grain size.

12. _____

13. Match each heat treatment listed with the corresponding statement.

_____ Thermal stress relieving

_____ Normalizing

_____ Annealing

_____ Spheroidizing

_____ Quench and temper

A. Quench rapidly to form martensite
B. Heat to 1000°F–1200°F (537°C–749°C)
C. Improves the ductility of high-carbon steel
D. Furnace cool to 50°F below the A_1 critical temperature
E. Cool in still air to room temperature

14. Give three reasons why a steel may be annealed. _____

15. Why is martensite tempered? _____

16. *True or False?* The final microstructure of a spheroidized high-carbon steel has small, separate spheres of ferrite in cementite.

16. _____

17. Which of the following will produce the maximum hardness in a steel?
    A. Heat to above the A_3 and slow cool
    B. Thermal stress-relieve the steel
    C. Heat to above the A_3 and rapidly quench
    D. Leave a steel in the "as welded" condition

17. _____

18. A part to be case-hardened is made out of _____ steel. The part is placed in a furnace that has a _____ atmosphere.

18. _____

19. *True or False?* White cast iron can be heat treated for 24 hours for each inch of thickness, and malleable cast iron will result.

19. _____

20. What happens to the alloys in aluminum after a solution heat treatment? _____

Job 29-1

LEARNING OBJECTIVE
★ In this job, you will practice tempering a cold chisel. This process is described under Heading 29.9.1 of the text. Also read Heading 29.6 of the text.

Name_____
Date _____ Class _____
Instructor _____
Grade _____

Safety Note: Always handle the chisel with metal tongs to prevent burns. Only when this job is completed, should you again touch the chisel with your hands. **Caution:** Always wear approved eye protection.

1. What is a typical carbon content of a cold chisel? 1. _____

2. Your instructor will provide a box with one or more cold chisels. Select one of the chisels. You will heat treat this chisel to obtain a hard cutting edge, hard and tough metal behind the cutting edge, and a hard and tough chisel body.

3. To what temperature should the chisel be heated initially? 3. _____

4. If a heat treating furnace is available, turn the furnace on and set the temperature to 30°F (15°C) above the answer in #3. (The temperature always varies slightly in a furnace, so set the furnace slightly high to make sure the metal goes above its critical temperature.)

5. The time that the chisel must be heated depends upon the thickness. Place the chisel in the furnace until the entire chisel is at the correct temperature.

6. If a heat treating furnace is not available, use an oxyacetylene torch. Before heating, apply temperature indicating crayons or liquids to the chisel, so you will know the temperature of the metal. You should apply two different temperature indicators, one for a temperature of 1350°F (732°C), the second for a temperature of 1400°F (760°C).

7. Heat the chisel to the correct temperature. Keep the chisel at this temperature so that the entire tool is at the correct temperature. The time depends on the thickness as described in #5 above.

8. After heating the chisel for the correct amount of time, place the cutting edge into a bucket of cold water. Only 1″ (25.4mm) of the chisel should be quenched in the water.

9. Leave the cutting edge in the water until the metal of the edge returns to a steel gray color.

10. Remove the chisel from the water.

11. Use a piece of emery cloth and polish the cutting edge of the chisel.

12. Heat from the rest of the chisel will begin to reheat the cutting edge.

13. Watch the polished surface as it changes color. Colors will progress from the body of the chisel down toward the cutting edge. Wait until the cutting edge turns completely purple. This is about 600°F (316°C).

14. The body of the chisel will have cooled to below the critical temperature. Place the entire chisel in the cold water. Leave it in the water for 60 seconds.

15. Remove the chisel. You have just quenched and tempered the cutting edge of the chisel. The edge is very hard. The body is hard and tough due to the small grains in the metal.

INSPECTION
Your instructor will check the condition of the chisel by using it. Neither the cutting edge nor the body should break.

Instructor's initials: _____

Lesson 30

INSPECTING AND TESTING WELDS

Name _____ Date _____ Lesson Grade _____
Class _____ Instructor _____

LEARNING OBJECTIVE
★ You will be able to explain the difference between a discontinuity and a defect. You will be able to use some nondestructive inspection methods to determine if a weld contains any defects. You will also be able to use some destructive tests to evaluate the properties of a weld.

INSTRUCTIONS
Carefully read Headings 30.1 through 30.19 of the text. Also study Figures 30-1 through 30-43 in the text. Then complete the following problems.

1. *True or False?* All welds have discontinuities. 1. _____

2. Define the term "discontinuity." _____
   _____
   _____

3. Which of the following statements is *false*? 3. _____
   A. All discontinuities are defects.
   B. All defects are discontinuities.
   C. NDE stands for nondestructive evaluation.
   D. NDT is a nonstandard term for NDE.

4. Match the following abbreviations with the correct test.

   MT   A. Leak test   4. A. _____
   ET   B. Visual test   B. _____
   RT
   LT   C. Liquid penetrant test   C. _____
   UT   D. Ultrasonic test   D. _____
   PT
   VT   E. Magnetic particle test   E. _____
           F. X-ray inspection   F. _____
           G. Eddy current inspection   G. _____

5. Magnetic particle inspection can only be used to check for flaws 5. _____
   at or near the _____. Magnetic particle inspection can be used
   only on materials that can be _____.   _____

Copyright Goodheart-Willcox Co., Inc.

359

6. In ultrasonic testing, an electronic device called a _____ is used to send _____ waves into the part to be tested.

6. _____
   _____

7. *True or False?* Radioactive isotopes or electronically produced X-rays can be used only to locate flaws at or near the surface of a metal being X-rayed.

7. _____

8. What can be determined from a tensile test? _____
_____
_____
_____

9. Label the parts of the hydraulic tensile test machine shown below.

   A. _____
   B. _____
   C. _____
   D. _____
   E. _____

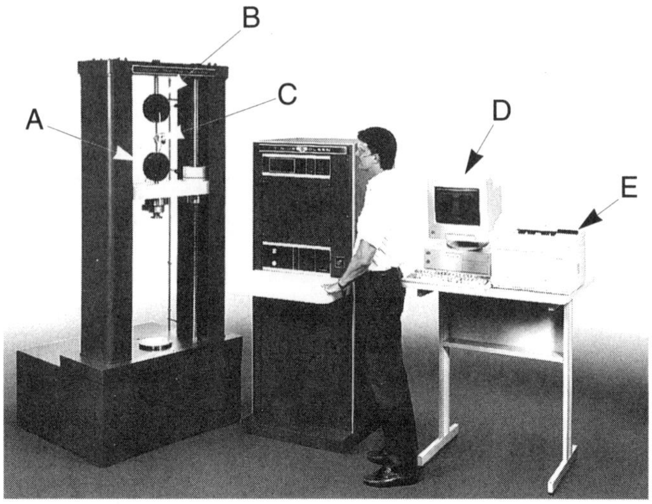

10. Match the following tests with the type of test or equipment that corresponds to that test. (The tests may be used more than once.)

    _____ Scleroscope
    _____ Quantitative analysis
    _____ Izod
    _____ Charpy
    _____ Brinell
    _____ Qualitative analysis

    A. Chemical test
    B. Impact test
    C. Hardness test

Job 30-1

LEARNING OBJECTIVE
* In this job, you will perform a magnetic particle inspection.

Name _____
Date _____ Class _____
Instructor _____
Grade _____

1. Obtain a piece of steel to be magnetic-particle tested. The steel may be one of your welds from another job or it may be a steel sample supplied by your instructor. Your instructor may have samples that are known to have cracks.

2. Examine the equipment. Does the equipment you will be using have a permanent magnet, an electromagnet or test prods? _____

3. What happens to the sides of a small crack when a magnetic field is applied? _____

4. What happens to the magnetic particles that are in the powder or liquid when a magnetic field is applied to the metal? _____

5. *True or False?* Magnetic particle testing can only be used on metals that can be magnetized.

5. _____

6. Label the parts of the magnetic particle test shown below.

   A. _____
   B. _____
   C. _____
   D. _____
   E. _____

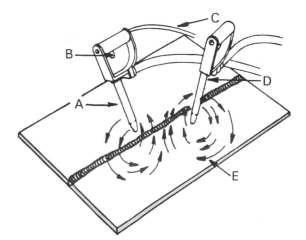

TEST PROCEDURE

7. Clean the surface to be tested, using the cleaner supplied.
8. Lightly sprinkle or dust the magnetic particles on the area to be tested. Do not apply the magnetic material too heavily, or the particles will not be able to move.
9. If a liquid is used, spray a light coating on the metal.
10. If using electrical prods or an electromagnet, turn on the power supply.
11. When using electrical prods, place the prods in contact with the metal. Then, push the control switch. Keep the prods in contact with the metal.
12. When using an electromagnet, place the yoke in contact with the metal and push the control switch. Keep the yoke in contact with the metal.
13. When using a permanent magnet, just place the magnet on the metal. No switch is used.
14. Whichever method is used, the results will be the same. The magnetic particles in the liquid or powder will gather around and point to any crack or flaw in the metal. Remember, even cracks or flaws that are below the surface (but very close to the surface) can be detected.
15. When using electrical prods or an electromagnet, it is often helpful to release the control switch so the magnetic field stops. Then, without moving the prods or yoke, again push the control switch. This will cause the magnetic particles to jump. They will move toward any crack or flaw.
16. Release the control switch and remove the prods or yoke.
17. Carefully examine the magnetic particles. They will gather around any flaw that can be detected. Remember, only flaws that are perpendicular to the magnetic field will be detected.
18. To finish the procedure, the part must be tested again. This time, the test prods or yoke must be 90° to the first test.
19. Completely inspect the part you are testing. Each area should be tested twice, once with yoke or prods in one direction, the second time with the yoke or prods 90° to the first.
20. Mark any flaws that you locate. You can mark them on the part or make a sketch of the part and mark them on the sketch. Your instructor will ask you to show how you located the flaws. Once you have located all the flaws, contact your instructor.

INSPECTION

Your instructor will ask you to demonstrate your magnetic particle inspection technique by checking a certain area of the part. Your instructor will examine the area to verify that the results you have are correct.

Instructor's initials: _____

Job 30-2

LEARNING OBJECTIVE
* In this job, you will perform a liquid penetrant inspection.

Name _____
Date _____ Class _____
Instructor _____
Grade _____

1. *True or False?* Liquid or dye penetrant inspection can detect flaws both at the surface and below the surface.

    1. _____

2. *True or False?* Dye penetrant inspection works only on non-magnetic metals.

    2. _____

3. A _____ is used to examine a part that has been tested using a fluorescent liquid.

    3. _____

4. Name the three ways excess penetrant can be removed. _____
_____
_____
_____

5. Obtain a part to be inspected. The part may be a weld from one of your other job assignments, or it may be a part supplied by your instructor. Your instructor may have parts that are known to have discontinuities (flaws) in them.

6. Clean the surface of the part with the cleaner provided.

7. Apply the penetrant to the surface. Wait 10 minutes so the penetrant can get into any cracks, pits or voids.

8. Remove the excess penetrant. (Use the proper solution as listed, because this could remove the penetrant from the flaws you are attempting to locate.)

9. Apply the developer. Wait 10 minutes so that any penetrant in a flaw will have time to be drawn out.

10. After waiting the recommended period, inspect the part. Dyes will be visible in regular light. A fluorescent penetrant must be inspected using a light as listed in question #3.

11. Record the locations of any flaws that you discover.

12. Clean the part after you have inspected it.

13. You may want to repeat the test on another piece of metal for practice.

INSPECTION
Your instructor may ask you to test another piece of material and locate all defects on it or to repeat the test on the same piece on which you practiced. Perform the dye penetrant test as described above. Carefully examine the part after the developer has been applied. Evaluate every indication and determine if it is a flaw. Many flaws, even very small ones, can be detected using penetrant inspection.

Instructor's initials: _____

Job 30-3

LEARNING OBJECTIVE
* In this job, you will perform a guided bend test. Read Headings 30.10, 31.5.2, and 31.5.3 of the text. Study Figures 30-9, 30-10, 30-22, and 30-23 in the text.

Name _____
Date _____ Class _____
Instructor_____
Grade _____

1. Obtain one or more groove-welded plates. The plates may be of any thickness. The plates should be at least 6″ (152mm) long.

2. If the plates you will bend are 3/4″ (19mm) thick or less, a face and root bend will be performed. If the plates are over 3/4″ (19mm) thick, side bends will be performed.

3. The radius to which the plates will be bent will be determined by your instructor.

4. If the plates are 3/4″ (19mm) thick or less, flame cut or shear two pieces 1.5″ (39mm) wide as shown below.

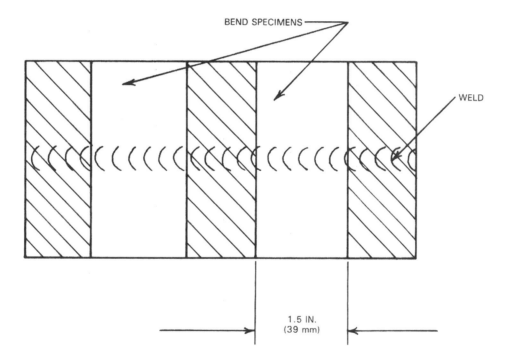

5. If the plates are over 3/4″ (19mm) thick, flame cut four pieces 1/2″ (13mm) wide.

6. Use a power grinder or a draw file and put a slight radius on the edge of the bend specimens. See the sketch below.

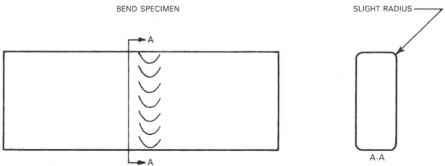

7. Your instructor will bend one of your specimens as a face bend. The other specimen will be bent as a root bend. If your plate is over 3/4″ (19mm), all samples will be bent as side bends.

Caution: Always wear approved eye protection when bend testing.

INSPECTION

Examine the bent specimens. There should not be any cracks or other defects over 1/8″ (3.2mm) long on the surface. If a defect begins from the edge, it can be ignored only if there is no evidence that it is due to poor welding techniques.

Instructor's initials: _____

Job 30-4

LEARNING OBJECTIVE
★ In this job, you will prepare samples for tensile testing and will perform a tensile test on the samples. Study Headings 30.13 and 31.5.1 of the text. Also study Figures 20-25, 30-26, and 31-9 in the text.

Name _____
Date _____ Class _____
Instructor _____
Grade _____

1. Obtain a groove-welded plate that is to be tensile tested. The sample can be one that you welded in one of the other job assignments. The sample should be about 6″ × 6″ (152mm × 152mm).

2. You will need to prepare a reduced section tension specimen. A reduced section specimen is shown in Figure 31-9 in the text.

3. If the metal you will be testing is steel, you can flame cut the specimen. If the metal is not steel, you will need to saw or shear the metal and then have the reduced section machined.

4. Your instructor may have a template to follow when flame cutting. If a template is not available, mark your plate using chalk or soapstone, then flame cut the correct shape as shown.

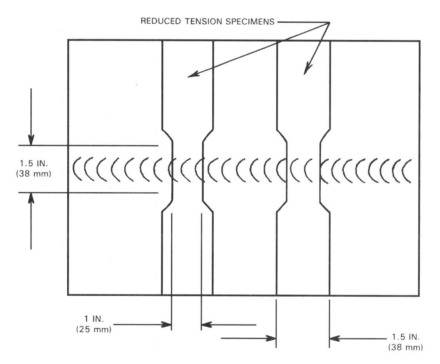

5. Grind the welds flush with the base metal on both sides of the plate.

6. Use a power grinder or draw file and break the sharp edges on the plate. Grind or file any rough areas until they are smooth. **Caution:** Wear approved eye protection when grinding.

7. Measure the width and thickness of your samples. Enter the measured values here.

	Sample #1	**Sample #2**
Width of sample	_____	_____
Thickness	_____	_____

8. Bring your prepared samples to your instructor for testing.

9. Your instructor will put your reduced section specimen in a tensile testing machine and apply a load to stretch the sample. Watch as the load (force) increases. Wear eye protection.

10. When your sample breaks, write down the load that it takes to break the sample.

    Load to break sample #1 _____

    Load to break sample #2 _____

11. Examine the broken samples with your instructor. Where did the break occur? Was it in the weld metal, heat-affected zone, or in the base metal?

    Sample #1 _____

    Sample #2 _____

12. Calculate the tensile strength of your sample. The formula to calculate the tensile strength is shown here. The formula is also shown under Heading 30.11 of the text and an example calculation is given there.

    $$\text{Tensile strength} = \frac{\text{tensile load}}{\text{area}} = \frac{\text{tensile load}}{\text{width} \times \text{thickness}}$$

    Calculate the tensile strength of your sample and show your work below.

    Tensile strength #1 = _____

    Tensile strength #2 = _____

INSPECTION

The tensile strength of your samples should be equal to or exceed the tensile strength of the base metal. If the tensile strength of your samples is considerably less than the base metal, this indicates a problem in the welding technique. Examine the fractured edge to make sure there is no evidence of slag, inclusions, pits, or voids. If you have slag or other defects in your weld, but the tensile strength of your sample was equal to or exceeded the tensile strength of your base metal, your welds have good strength. However, you should practice to eliminate these defects from your welding.

Instructor's initials: _____

Job 30-5

LEARNING OBJECTIVE
* In this job, you will test the hardness of different metals.

Name _____
Date _____ Class _____
Instructor _____
Grade _____

1. Define *hardness*. _____
   _____

2. A _____ is *not* a hardness test.
   A. Rockwell test
   B. Charpy test
   C. Brinell test
   E. Scleroscope test

   2. _____

3. When using a Rockwell hardness test, _____ loads are applied to the sample.

   3. _____

4. Which tests require the operator to use a microscope to measure the indentation made?

   4. _____

5. Label the parts of the hardness tester shown below.
   A. _____
   B. _____
   C. _____
   D. _____
   E. _____
   F. _____

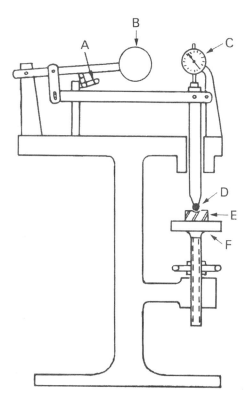

6. Obtain four samples on which to perform hardness tests. The samples will be provided by your instructor. The samples may be low-carbon steel, high-carbon steel, copper, aluminum, or other metals. The samples may be the same metal, but have different heat treatments.

7. The Rockwell B scale is used on softer metals like aluminum and copper. The Rockwell C scale is used on harder metals. The Brinell and Scleroscope scales cover a large range from soft to hard metals.

8. First, the testing equipment must be calibrated. Samples with a known hardness are supplied with the testing equipment. Test the hardness of a sample and adjust the equipment so it reads the hardness correctly. It is recommended that you recalibrate the equipment every time you use it. You should also recalibrate it for each metal type you will be testing.

9. Test the hardness of the sample you have. You should take three hardness readings on each sample.

10. Record the hardness value for each reading you take. Average the three values and record the average below. Look at the three hardness readings. They should all be close together. If one reading is very high or very low, do not include it in your average.

SAMPLE

Hardness Reading	#1	#2	#3	#4
#1				
#2				
#3				
Average				

INSPECTION
Your instructor will examine the three different hardness readings and the average. Your instructor will compare the average you obtained to the correct hardness value for that metal.

Instructor's initials: _____

Lesson 31

PROCEDURE AND WELDER QUALIFICATIONS

Name _____ Date _____ Lesson Grade _____
Class _____ Instructor _____

LEARNING OBJECTIVE

★ You will be able to use different codes, specifications, and standards to determine how welding will be performed. You will also be able to weld in accordance with a code or standard.

INSTRUCTIONS

Carefully read Headings 31.1 through 31.5.5 of the text. Also study Figures 31-1 through 31-15 in the text. Then complete the following problems.

1. Write the following abbreviations in the blank provided with the corresponding agency, association, or society.

   MIL   A. Federal Aviation Administration   1. A. _____
   AWS   B. American Society of Mechanical Engineers   B. _____
   FAA   C. American Society for Testing and Materials C. _____
   ASTM   D. Military Specifications   D. _____
   ASME   E. Interstate Commerce Commission   E. _____
   ICC   F. American Welding Society   F. _____

2. What is covered in ASME Section IX? _____
   _____

3. When must a code or specification be followed? _____
   _____

4. *True or False?* Every WPS must have a WPQR to document the   4. _____
   quality of the welds produced using that procedure.

5. Which of the following are essential variables as listed on a   5. _____
   Welding Procedure Specification?
   A. The base metals to be welded
   B. The filler metal to be used
   C. The position in which the welding will be done
   D. The thickness of the plates to be welded
   E. Type of groove
   F. Postheat temperature

Copyright Goodheart-Willcox Co., Inc.

371

6. *True or False?* Whenever there is a change in an essential or a nonessential variable, the WPS must be requalified.

6. _____

7. Every welder must qualify to each welding _____ specification to which he or she will be required to weld.

7. _____

8. A welder performs a welding performance qualification test on 1″ (25.4mm) thick plate. Use Figure 31-3, titled Tension Test and Transverse Bend Tests, in the text.

   A. On what range of base metal is the welder qualified to weld?

   B. How many tension specimens will be tested?

   C. How many side bends will be tested?

   8. A. _____

   B. _____

   C. _____

9. A welder may be required to requalify for these two reasons: _____

   _____

   _____

   _____

   _____

10. Label the different positions shown with the number and letter combination used to designate that welding position.

    A. _____
    B. _____
    C. _____
    D. _____
    E. _____
    F. _____

Job 31-1

LEARNING OBJECTIVE

★ In this job, you will establish and write a welding procedure specification (WPS) for welding a groove weld. You will attempt to qualify the WPS that you establish. You will remove samples that will be tested to determine if you qualify to weld to the WPS. Welding will be done in the flat welding position.

Name _____
Date _____ Class _____
Instructor _____
Grade _____

NOTE TO THE STUDENT

The values that you enter in the WPS are to be determined by you. Each welder knows what electrode, what current, and what groove type will produce the best weld for him or her. From your knowledge of welding, determine for yourself what values will produce the best results. Enter these values in the WPS. When working for a welding company that welds per codes and specifications, you will have to weld to the conditions set up on their welding procedure specifications.

If you are not sure what values to enter on the WPS, obtain two or more pieces of metal that your instructor assigns and weld them together using the process your instructor selects. Pay attention to the parameters you use. After experimenting on practice pieces, enter the values that work best on the WPS.

1. You will be required to fill out all the information on the welding procedure specification form prior to doing any welding.

2. Your instructor will select what process, what material, and what material thickness will be used for this assignment.

3. If you are using an oxyfuel gas process, enter the fuel and oxygen flow rates. If you are using a process that requires gas shielding, enter the type of gas shielding you will be using and the flow rate.

4. If you are using an oxyfuel gas process, enter the filler rod type and diameter you will be using.

5. If you are using an arc welding process, enter the electrode type and the diameter. Notice that a different type or diameter can be used for each pass. Enter the type and diameter you will use for each pass.

6. If you are using an arc process, enter the current, voltage, polarity, and travel speed as required for the process you are using.

7. Enter the type of groove you will be welding and the dimensions required on the WPS.

8. If you will be using any preheat or postweld heat treatment, enter the appropriate temperatures.

9. Enter the type of groove you will be welding and the dimension required on the WPS.

10. If any spaces are left blank, write "N/A" for not applicable.

11. Show your completed WPS to your instructor.

12. After reviewing your WPS, your instructor will give you two pieces of the metal you are to weld. The samples are to be 8″ × 3″ (203mm × 67mm) each.

13. Prepare the metal as stated on your WPS.

14. Obtain the type of electrode or filler metal you will use to weld.

15. Set up your welding station. Keep in mind all safety precautions that you set on your WPS.

16. Set up the welding equipment. Adjust or set it to provide the parameters that you set on your WPS.
17. Tack weld the two pieces together.
18. When you are ready to weld, contact your instructor. Your instructor will review your machine setup and may watch you as you weld to make sure you are meeting the parameters you listed on your WPS.
19. From your completed weld, cut two 1.5″ (39mm) wide specimens for bend testing. Also cut one 1.5″ (39mm) wide specimen for tensile testing. The location of each sample is shown in the figure below.

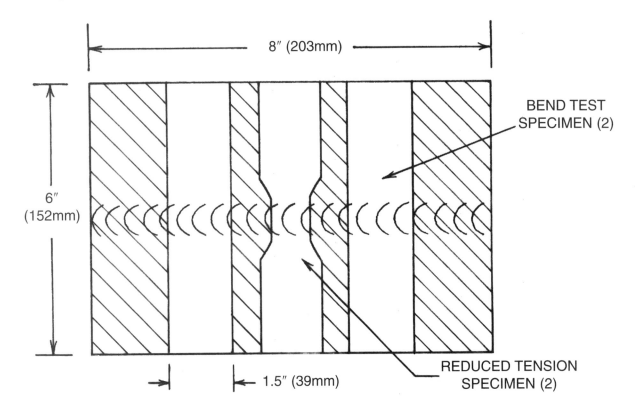

20. Prepare the two bend test samples as described in Job 30-3.
21. Prepare one reduced tension specimen as described in Job 30-4.
22. Record the width and thickness of your reduced tension specimen on the welding procedure qualification record (WPQR).
23. Bring the tensile and bend specimens to your instructor for testing. Wear approved eye protection when testing.
24. After bend testing, have your instructor check "pass" or "fail" on the WPQR. After performing the tensile test, calculate the tensile strength of the tensile specimen and record the value on the WPQR.

INSPECTION
Your instructor will review the results of the tensile test and bend tests to see if your welding is satisfactory. Your instructor will review your welding technique and the results of the destructive tests. Your instructor will then grade your ability to produce a satisfactory weld.

Instructor's initials: _____

Welding Procedure Specification

Name _____ Date _____ Grade _____
Class _____ Instructor _____

Base metal type _____

Thickness _____

Process _____

Gas

Fuel gas flow _____

Oxygen flow _____

Shielding gas composition _____

Shielding gas flow _____

Electrode

	Root Pass	2nd Pass	Fill Passes	Cover Pass
Electrode type	_____	_____	_____	_____
Electrode diameter	_____	_____	_____	_____
Filler metal type	_____	_____	_____	_____
Filler metal diameter	_____	_____	_____	_____

Welding Parameters

	Root Pass	2nd Pass	Fill Passes	Cover Pass
Voltage	_____	_____	_____	_____
Current	_____	_____	_____	_____
Polarity	_____	_____	_____	_____
Travel speed	_____	_____	_____	_____
Contact tube to work distance*	_____	_____	_____	_____

Type of Groove

Type of groove _____

Root gap _____

Included angle _____

Root face _____

Heat Treatment

Preheat temperature _____

Postweld heat treatment temperature _____

*Used on automatic welding only

Welding Procedure Qualification Record (WPQR)

Tension Test

Prepare the reduced tension specimen as described in Job 28-4.

Width of specimen   _____

Thickness of specimen   _____

Load to break specimen   _____

Tensile strength   _____

Show your calculations.

Bend Test

Prepare the bend test specimens as described in Job 30-3.

Examine bend specimens

	Pass	Fail
Face Bend	( )	( )
Root Bend	( )	( )

Job 31-2

LEARNING OBJECTIVE
* In this job, you will perform a fillet weld performance test. Your instructor will select the type of material to be used and the process to be used.

Name _____
Date _____ Class _____
Instructor _____
Grade _____

1. Obtain two pieces of material. One piece should be 6" × 4" × 3/16" (152mm × 104mm × 4.8mm), the other piece should be 6" × 3" × 3/16" (152mm × 76mm × 4.8mm).

2. Obtain the correct electrode or filler metal to be used on the base metal you will be welding.

3. Set up the welding equipment as described in the text.

4. Align the pieces to be welded as shown in the figure below.

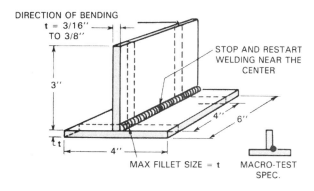

5. Tack weld the plates into this position.

6. Weld from one end toward the center. When you reach the center, stop welding. Restart your weld at the center and finish the fillet weld.

7. Clamp the base piece in a vise. Apply a load, using a long bar or other tool, so the vertical piece folds down over the weld. Continue to bend the vertical piece until the weld breaks or the two pieces touch one another. **Caution:** Use approved eye protection while performing this test.

INSPECTION
Examine the exposed surface of the weld and look for any evidence of lack of fusion, inclusions, pits, or other discontinuities. There should be very little or no evidence of these flaws.

Instructor's initials: _____

Lesson 32
THE WELDING SHOP

Name _____ Date _____ Lesson Grade _____
Class _____ Instructor _____

LEARNING OBJECTIVE
* You will be able to identify and use various equipment in a welding shop.

INSTRUCTIONS
Read Headings 32.1 through 32.15 of the text. Also study Figures 32-1 through 32-37 in the text. Then complete the following problems.

1. Which of the following is *not* a reason to use a ventilation system?   1. _____
   A. Supply clean air and oxygen to workers
   B. Keep gases and fumes to a minimum
   C. Release dust particles into the air
   D. Heat the replacement air during cold weather

2. A _____ shear is used for cutting thin metals.   2. _____

3. Metal must often be formed. What machine is used to bend most metal? _____
   _____

4. List the three main types of preheaters. _____
   _____
   _____
   _____

5. The most important reason why jigs and fixtures are used in a   5. _____
   welding shop is to _____.
   A add additional heat to the parts being welded
   B. make the welding easier for the welders
   C. reduce spatter
   D. prevent the parts from warping or buckling during welding

6. *True or False?* Air pressure that varies from 20 psig to 100 psig   6. _____
   (138 kPa to 689 kPa) is used to propel abrasives during a blast
   cleaning operation.

7. What piece of equipment is shown below?   7. _____

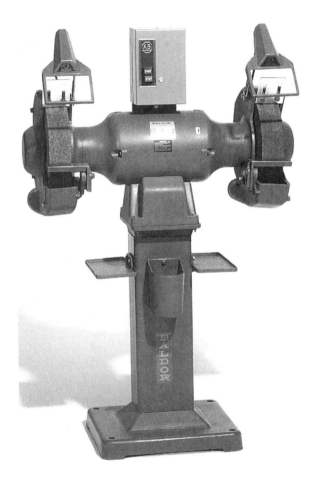

8. From the list of welding shop tools listed in Heading 32.11 of the text, think about each tool and list the six that you think would be used most often by a welder in industry. _____
   _____
   _____
   _____

9. _____ requires little or no equipment.   9. _____
   A. Visual inspection
   B. X-ray inspection
   C. Hardness testing
   D. Bend testing

10. The job of an _____ is vital to a shop's financial success.   10. _____

Lesson 33

GETTING AND HOLDING A JOB IN THE WELDING INDUSTRY

Name _____ Date _____ Lesson Grade _____
Class _____ Instructor _____

> **LEARNING OBJECTIVE**
> ★ You will be able to identify the information that you will need to prepare for a job interview and to complete a job application. You will also be able to list the attitudes that you should display during an interview and on the job.
>
> **INSTRUCTIONS**
> Read Headings 33.1 through 33.9 in the text. Then complete the following problems.

1. Assume that you are applying for a job as a gas metal arc welder in a large steel fabrication plant. Complete, to the best of your ability, the blank job application #1 that is part of this lesson.

2. After completing the application, list all the items for which you did not have sufficient information. These and other items of information you should have written on a piece of paper that you can take with you when you apply for a job. _____

    _____
    _____
    _____
    _____

3. How do you intend to enhance your personal appearance in preparation to job seeking and for a job interview?

    _____
    _____
    _____

4. If asked what you hope to be doing three years from now, what would you answer? This shows an employer that you have goals. _____

    _____
    _____
    _____

Copyright Goodheart-Willcox Co., Inc.

381

5. If you were applying for a welder's job, what would you answer if asked, "Why do you want to work for our company?" _____
   _____
   _____
   _____
   _____

6. Imagine that you are an employer. What would you think of someone you were interviewing, if they kept looking at the floor or couldn't look you square in the eye? _____
   _____
   _____
   _____
   _____

7. If you did poorly in completing this lesson or the job application, ask yourself the following questions.

   A. "Am I able to follow written directions?" (Yes/No)   7. A. _____

   B. "Am I able to express myself in writing?" (Yes/No)   B. _____

   If your answers to either of these is "No," then you should work to gain these skills.

Name: _____   Lesson 33   Getting and Holding a Job in the Welding Industry   383

APPLICATION FOR EMPLOYMENT

Information furnished on this application is subject to verification. Misrepresentation of data could result in rejection as an applicant or subsequent dismissal if employed.

PERSONAL DATA

Last Name _____ First _____ M.I. _____ Social Security Number _____ Are you under 18? ___ Yes ___ No

Address (Number and street) _____ City _____ State _____ Zip Code _____ County _____

Permanent Address (If other than above) _____ City _____ State _____ Zip Code _____ County _____

Phone (area code and number) Best time to call Day (circle) | Do you have invitations to interview at other locations in the company? ___ Yes ___ No
( ) _____ _____ M T W Th F | If yes, specify: _____

Phone (alternate number)
( ) _____ _____ M T W Th F

Only U.S. citizens or aliens who have a legal right to work in the United States are eligible for employment.

Are you a U.S. citizen or permanent resident alien? ___ Yes ___ No

If no, what is your visa status? _____

If alien, what is your registration number? _____

Have you previously applied at or worked for this company?

Applied: ___ Yes ___ No   Dates: _____   Location: _____
Worked: ___ Yes ___ No   Dates: _____   Location: _____

Are you willing to relocate? ___ Yes ___ No   Geographic preference: _____

Geographic limitations: _____   Are you willing to travel as required? ___ Yes ___ No

Are you willing to work overtime as required? ___ Yes ___ No

Are you willing to work shifts as required? ___ Yes ___ No

Are you willing to take a drug test? ___ Yes ___ No

An Equal Opportunity Employer

Copyright Goodheart-Willcox Co., Inc.

Date available for employment _____ Total annual salary requirement _____

What led you to apply here? (check one)

Own initiative:
___ Walk-in
___ Write-in

Answered ad:
___ Walk-in
___ Write-in

Referred by:
___ Private empl. agency
___ State empl. agency
___ Company employee
___ Campus recruited or referral

___ Professional organization
___ Other (explain) _____

AREAS OF WORK INTEREST

Type of work desired (check one): ___ Full-time ___ Part-time ___ Temporary ___ Summer

Position desired: _____

Check your primary field of work interest.

Professional/Technical
___ Administrative
___ Accounting
___ Computer Systems/Programming
___ Finance
___ Human Resources
___ Labor Relations
___ Legal
___ Manufacturing Engineering
___ Plant Engineering
___ Product Engineering
___ Production Control
___ Public Relations
___ Purchasing
___ Quality Control
___ Research & Development
___ Sales/Marketing
___ Transportation/Traffic

Office/Clerical
___ General Clerk
___ Mail Clerk
___ Secretary
___ Typist
___ Word Processor

Skilled/Production/Maintenance Work
___ Mechanic/Repairer
___ Machine Operation
___ Material Handling
___ Maintenance Work
___ Other

EDUCATION

Name, city, and state of educational institution	Graduated? Yes / No	If no degree, credits earned	Type of degree rec'd. or expected Mo. / Yr.	Major / Minor	Grade Point Maj. / Cum. / Base	Class standing by quarter Top 2 3 4
High School						
Technical/ GED/Other						
College or University						

What foreign languages do you: Read _____ Speak _____ Write _____

Did you work during your educational program? ___ Yes ___ No — Average hours per week: _____

Copyright Goodheart-Willcox Co., Inc.

WORK EXPERIENCE

Start with present or most recent position. You may include military service, summer positions, and volunteer work experience. (You may attach a separate page.)

May we contact your present employer? _____ Yes _____ No

Employer (present or most recent) _____ Address _____

Dates employed
From Mo. _____ Yr. _____ To Mo. _____ Yr. _____

Position title _____

Number of persons supervised _____

Starting base pay $ _____ per _____
Additional compensation $ _____

Ending base pay $ _____ per _____
Additional compensation $ _____

Duties _____

Average number of work hours per week _____ Reason for leaving _____

Employer _____ Address _____

Dates employed
From Mo. _____ Yr. _____ To Mo. _____ Yr. _____

Position title _____

Number of persons supervised _____

Starting base pay $ _____ per _____
Additional compensation $ _____

Ending base pay $ _____ per _____
Additional compensation $ _____

Duties _____

Average number of work hours per week _____ Reason for leaving _____

WORK EXPERIENCE *(Continued)*

Employer _____ Address _____

Dates employed
From Mo. ____ Yr. ____ To Mo. ____ Yr. ____

Position title _____

Number of persons supervised _____

Starting base pay
$ _____ per _____
Additional compensation
$ _____

Ending base pay
$ _____ per _____
Additional compensation
$ _____

Duties _____

Average number of work hours per week _____ Reason for leaving _____

ACTIVITIES AND ACHIEVEMENTS

(You may exclude those which indicate race, color, religion, sex, marital status, age, or national origin, handicapped or veteran status.)

Honors (include societies and scholarships) _____

Publications _____

Professional and technical associations _____

Patents _____

TO BE COMPLETED BY OFFICE AND CLERICAL APPLICANTS ONLY

Steno/clerical skills (words per minute)

Typing speed _____ Shorthand speed _____ Other related skills _____

What office machines do you operate, e.g., word processor, keypunch, etc.? _____

CAREFULLY READ THE PARAGRAPHS BELOW BEFORE SIGNING AND DATING THE APPLICATION

I acknowledge that the information I have furnished is correct to the best of my knowledge and understand that falsification of this information could be grounds for disciplinary action to include retraction of employment offer or dismissal if employment has taken place.

I understand that I shall not become an employee of this company or any of its subsidiaries until I have signed an employment agreement with final approval of the company and that such employment will be contingent upon my submitting to a physical examination to assure that I am physically able to perform the type of employment for which I am applying and such employment will be subject to verification of the information I have provided on this application and any related documents or resume.

I understand that it may be necessary for the company to use consumer reporting agencies to verify such information provided by me and that any information obtained may be retained by those agencies and subsequently used in later reports. I understand that the company chooses consumer reporting agencies with care. I understand that should use of a consumer reporting agency for verification of the information provided by me be deemed necessary, I will be contacted to complete releases to the parties from whom verification will be sought, and that I may make a written request to the company location to which I applied for a copy of the verification report. Alternatively, I may make a written request of the consumer reporting agencies for additional information as to the nature and scope of the verification report. I understand that if the need to correct or amend any statement on the verification report arises, I may notify the company location to which I made application. I understand that the company will not make any unwarranted disclosure of information generated in the process of application to persons outside the company without my consent. I understand that if I am selected for employment, evidence of U.S. citizenship or U.S. permanent resident status must be provided to the company.

Signature _____ Date _____

WE APPRECIATE YOUR INTEREST IN THE COMPANY AND THE TIME YOU HAVE TAKEN TO PREPARE THIS APPLICATION.

DO NOT COMPLETE — FOR PERSONNEL USE ONLY

Tests	Test results Pass / Fail	Tests	Test results Pass / Fail
Typing		OBK-LMR	
Stenographic dictation		MIS-C	
		TBJ-P	

Test administrator _____ Date _____

CONVICTION DATA SHEET

This portion of your application **must** be completed, but will be detached and maintained separately. It will be used only when the information is relevant to your application. If employed, this information will not become part of your permanent employment record.

Last Name _____ First _____ M.I. _____ Social Security Number _____

Have you ever been convicted of a felony or convicted of a lesser crime within the last five years? Do not include arrests with conviction, charges expunged, convictions adjudged "youthful offender" or "juvenile," or convictions for minor traffic violations.   Yes _____   No _____

If yes, please briefly describe the circumstances of your conviction and your name at that time; indicating the date, nature, and place of the offense and disposition of the case including any rehabilitation. Your answer is looked upon as only one of the factors considered in the employment decision and is evaluated in terms of the nature, severity, and date of the offense.

_____

_____

Signature of applicant _____   Date _____

SUPPLEMENTAL DATA SHEET

The following applicant information is requested for the purpose of preparing periodic reports to the government or other recordkeeping in connection with government requirements. We encourage you to complete this section, but your employment prospects will not be adversely affected should you choose not to provide this information. This portion of the employment application will not become part of your application or employment file.

Last Name _____ First _____ M.I. _____ Social Security Number _____

Are you a disabled veteran?   Yes _____   No _____

Are you a Vietnam Era veteran?   Yes _____   No _____

Do you have a physical or mental handicap?   Yes _____   No _____   If yes, please describe: _____

Sex/Race — Ethnic (check one)

_____ White male
_____ White female
_____ Hispanic male
_____ Hispanic female
_____ Black male
_____ Black female
_____ American Indian or Alaskan Native male
_____ American Indian or Alaskan Native female
_____ Asian or Pacific Islander male
_____ Asian or Pacific Islander female
_____ Other _____

Name: _____ Lesson 33 Getting and Holding a Job in the Welding Industry 389

Job 33-1

> **LEARNING OBJECTIVE**
> ✳ In this job, you will be asked to complete a job application. This application should be completed in class while under your instructor's supervision. You may use any information that you have prepared and have with you regarding addresses, social security number, telephone number, etc.

Please print all information. You must fully and accurately complete the application.

APPLICATION FOR EMPLOYMENT

PERSONAL INFORMATION

Date _____ Social Security Number _____

Name _____
    Last   First   Middle

Present Address _____
    Street   City   State

Permanent Address _____
    Street   City   State

Phone No. _____ Alternate No. _____

If related to anyone in our employ, state name and department _____ Referred by _____

EMPLOYMENT DESIRED

Position _____ Date you can start _____ Salary desired _____

Are you employed now? _____ If so may we inquire of your present employer? _____

Ever applied to this company before? _____ Where _____ When _____

EDUCATION

	Name and Location of School	Years Completed	Subjects Studied	Degree Earned
Grammar School				
High School				
College/University				
Trade, Business or Correspondence School				

Subject of special study or research work _____

Copyright Goodheart-Willcox Co., Inc.

What foreign languages do you speak fluently? _____ Read fluently? _____ Write fluently? _____

U.S. Military service _____ Rank _____ Present membership in National Guard or Reserves _____

Activities other than religious (Exclude organizations the name or character of which indicates the race, creed, color or national origin of its members.) _____

FORMER EMPLOYERS List below last three employers starting with last one first

Date Month and Year	Name and Address of Employer	Salary	Position	Reason for Leaving
From				
To				
From				
To				
From				
To				

REFERENCES Give below the names of two persons not related to you whom you have known for at least one year

	Name	Address	Job Title	Years Acquainted
1				
2				

PHYSICAL RECORD

Have you any disabilities that might affect your ability to perform this job? _____

In case of emergency notify _____
Name   Address   Phone No.

We are an equal opportunity employer. We are dedicated to a policy of non-discrimination in employment on any basis including race, creed, age, sex, religion, national origin, height, weight, marital status, or disability.

I understand that to accept employment, I must be lawfully authorized to work in the United States, and I must present documents to prove my eligibility.

I understand that the company may thoroughly investigate my work and personal history and verify all data given on this application, on related papers, and in interviews. I authorize all individuals, schools, and firms named therein, except my current employer if so noted, to provide any information requested about me, and I release them from all liability for damage in providing this information.

The information on the I application and any made in conjunction with this application is correct and true to the best of my knowledge. I understand that any false or misleading statement made by me in connection with this application or the failure to disclose any material will be grounds for immediate dismissal.

In consideration of my employment, I agree to conform to the rules and regulations of this company, and my employment and compensation can be terminated, with or without notice, at any time, at the option of either the company or myself. I understand that no manager or representative of the company, other than the president/owner of the company, has any authority to enter into any agreement for employment for any specified period of time, or to make any agreement contrary to the foregoing.

I authorize investigation of all statements contained in this application. I understand that misrepresentation or omission of facts called for is cause for dismissal.

Date _____   Signature _____